Design for Static Mechanical Strength

A monograph covering:
- Stress and Strain
- Theories of Failure
- Factors of Safety
- Stress Concentrations
- Material Properties
- Metal Processing

Carl F. Zorowski

Design for Static Mechanical Strength

Copyright 2016
All Rights Reserved

Design for Static Mechanical Strength

Table of Contents

Chapter 1 – Stress and Strain
- Definition of Stress — 1
- Two Dimensional Stress State — 2
- Rotated Stress State — 3
- Mohr's Circle Construction — 3
- Principal Stresses — 4
- Special Cases — 6
- Strain Components — 7
- Principal Strains — 8
- Three D Stress/ Strain Components — 9
- Generalized Hook's Law — 10
- A Special Property — 11
- Plane Stress State — 12
- Plane Strain State — 13
- Principal Stress Equation — 15
- Mohr's Circle – 3 D — 16
- Example Problem — 17

Chapter 2 – Theories of Static Failure
- Designer's Dilemma — 21
- Theories of Static Failure — 22
- Maximum Stress Theory — 23
- Maximum Shear Stress Theory — 24
- Yielding in Pure Shear — 26
- Distortion Energy Theory — 27
- Comparison of Ductile Theories — 30
- Applicable Ductile Theories — 31
- Ductile Theory Example — 31

Design for Static Mechanical Strength

- Mohr's Theory   35
- Coulomb Mohr Theory   37
- Modified Coulomb Mohr Theory   38
- Applicable Brittle Theories   39
- Brittle Theory Example   40

Chapter 3 – Factors of Safety
- Definition   45
- Variation Scenario   46
- Minimum Factor of Safety   48
- Max Shear Stress Theory   50
- Modified Theories of Failure   51
- Effect of Factor of Safety   52
- Statistical Factor of Safety   53
- Apply Gaussian Distribution   55
- Example Values of t_f 56
- Sample Problem   57
- Problem Solution   57
- Additional Approximations   58

Chapter 4 – Stress Concentrations
- Occurrence of Phenomena   61
- Flow Visualization   62
- Symbolic Representation   63
- A Theoretical Solution   64
- K_t variation with a/b   65
- Modeling Techniques   65
- Published Information   67
-

Design for Static Mechanical Strength

- Stress Concentration Factors 68
  (shoulder filet-axial load)
- Stress Concentration Factors 69
  (shoulder fillet – bending load)
- Stress Concentration Factors 69
  (shoulder filet – torsion load)
- Numerical Example 70
- Effect of Material Behavior 71
- Notch Sensitivity and Fatigue 72
- Sample Notch Sensitivity Chart 73
- General Guidelines 74

Chapter 5 – Material Properties
- Testing Overview 77
- Tensile Test 77
- Test Results Interpretation 78
- Compression Test 81
- Sample Properties 82
- Carbon Steel 82
- Impact Testing 84
- Hardness Teating 85
- Brinell Hardness 85
- Rockwell Hardness 86

Chapter 6 – Metal Processing
- Processing Overview 89
- Sand Casting 90
- Investment Casting 91
- Shell Casting 93
- Die Casting 94

Design for Static Mechanical Strength

- Rolling (ingot to billet)   96
- Rotary Piercing   97
- Extrusion   97
- Forging   99
- Cold and Hot Work Compared   100
- Some Cold Working Processes   101
- Heat Treatment   103
- Annealing   103
- Normalizing   104
- Quenching   105
- Tempering   106
- [1]Effects of Tempering   107
- Case Hardening   108

Preface

Design for Static Mechanical Strength is a collection of topics relevant to minimizing mechanical design failure under static loading. As a monograph it attempts to present the essential classic knowledge base of the subject together with a degree of modeling and analytic detail that is complete while exercising brevity. It's content includes stress and strain, theories of failure, factors of safety, stress concentrations and the effect of processing on metal properties.

The goal of this monograph is to provide an understanding of the basic theory and models that are appropriate to the engineering application of the relevant subject matter of static mechanical strength in a succinct manner. It is not intended to be a textbook or comprehensive reference source. Its purpose is to assist the once acquainted reader in recalling relevant content material with greater understanding or to provide a concise complimentary supplement to students acquiring the knowledge for the first time in a structured learning environment.

Of the one hundred plus pages of the monograph forty percent is dedicated to graphic representations and analytic developments. This emphasis on pictorial and mathematical content is deliberate to address the visual learning style of most engineers. The mathematical developments are principally algebraic and logically structured for ease of comprehension. When coupled with the accompanying text they provide for a richer and

Design for Static Mechanical Strength

deeper understanding of the fundamentals and principles presented. It is recommended that they be considered in this manner and not be overlooked. Mathematics as the language of engineering provides a unique basis for the knowledge that undergirds the practice of the profession.

The material contained in this monograph is abstracted from a course supplement (*Design for Strength and Endurance,* copyright-2002, ISBN 0-973126-1-3*)* available in PDF format in its entirety at www.designforstrength.com. This supplement was prepared for and used in a Mechanical Design Engineering distance education course taught by the author at North Carolina State University.

The value and benefit of this monograph as an effective teaching/learning instrument is left to the judgment of the user.

Carl F. Zorowski
Cary, NC
October 2016

Design for Static Mechanical Strength

Design for Static Mechanical Strength

Design for Static Mechanical Strength

Chapter 1 – Stress and Strain

Chapter 1 deals with the definition of stress and strain, two dimensional stress states, stresses and strains relative to rotated axes, Mohr's circle construction, principal normal stresses, maximum shear stress, generalized Hooke's law, plane stress and plane strain states, three dimensional principal stresses and an example application.

Definition of Stress

Consider a body in equilibrium under the action of forces F1 through F4. On a plane cut through the body depicted by the light gray dotted surface an element of area ΔA has acting on it a component of force ΔF as defined by equilibrium of that section of the body. The limit of the ratio of the normal component ΔF_n of ΔF over ΔA as ΔA approaches zero is defined as the normal stress component sigma, σ.

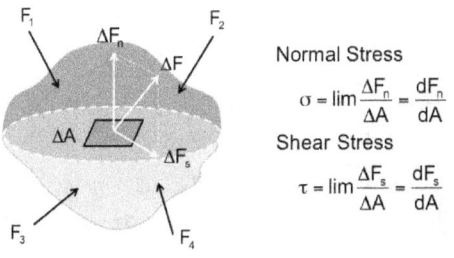

Normal Stress

$$\sigma = \lim \frac{\Delta F_n}{\Delta A} = \frac{dF_n}{dA}$$

Shear Stress

$$\tau = \lim \frac{\Delta F_s}{\Delta A} = \frac{dF_s}{dA}$$

Figure 1-1 Stress Definition

1

Design for Static Mechanical Strength

In a similar fashion the limit of the ratio of the horizontal component ΔF_s of ΔF over ΔA as ΔA approaches zero is defined as the shear stress tau, τ.

Two Dimensional Stress State

A two-dimensional stress state is defined by the stress components on the dx dz and dy dz faces of a cubical element of dimensions dx dy dz. A normal stress σ_x and a shear stress τ_{xy} act on the dy dz faces (left and right side) of the element. A normal stress σ_y and a shear stress τ_{yx} act on the dx dz (top and bottom) faces of the element (Figure 1-2).

The normal stresses create forces that are in equilibrium in the x and y directions. Moment equilibrium of the shear forces created by the shear stresses requires that τ_{xy} be equal to τ_{yx}. Hence the two dimensional state consists of three components of stress, σ_x, σ_y and τ_{xy}.

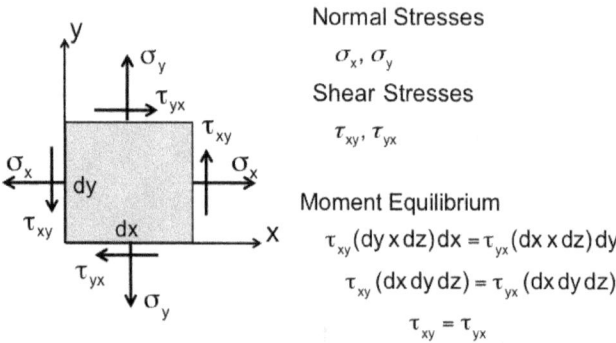

Figure 1-2 Two Dimensional Stress State

Design for Static Mechanical Strength

Rotated Stress State

If the element is rotated through an angle θ the stresses σ_x', σ_y' and τ_{xy}' relative to the new x' y' coordinate system are given by the three equations listed in Figure 1-3. σ_x' and σ_y' are similar in format as related to the trigonometric functions with the exception of the negative τ_{xy} term in σ_y'.

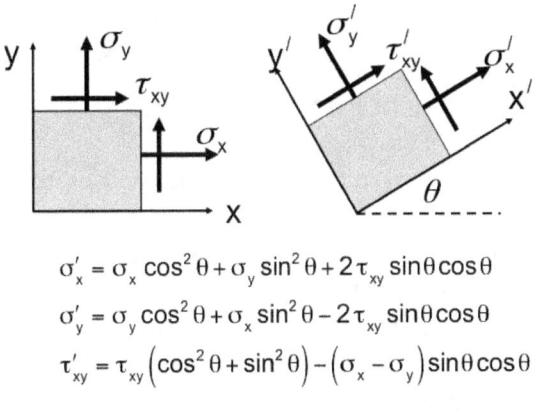

$$\sigma_x' = \sigma_x \cos^2\theta + \sigma_y \sin^2\theta + 2\tau_{xy} \sin\theta\cos\theta$$

$$\sigma_y' = \sigma_y \cos^2\theta + \sigma_x \sin^2\theta - 2\tau_{xy} \sin\theta\cos\theta$$

$$\tau_{xy}' = \tau_{xy}\left(\cos^2\theta + \sin^2\theta\right) - \left(\sigma_x - \sigma_y\right)\sin\theta\cos\theta$$

Figure 1-3 Rotated Stress State

The effect of these equations on the rotated stress components as a function of the angle θ can best interpreted by their graphical representation.

Mohr's Circle Construction

The two coordinates points σ_x, τ_{xy} and σ_y, -τ_{xy} are plotted on a sigma, σ, / tau, τ, coordinate axis system in Figure 1-4. A line drawn between these two points defines the diameter of what is called Mohr's circle.

Design for Static Mechanical Strength

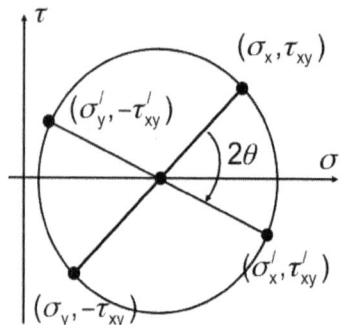

Figure 1-4 Mohr Circle Construction

The stresses σ_x' σ_y' and τ_{xy}' as defined by the equations in Figure 1-3 are represented by the end points of a diameter rotated an angle 2θ clockwise from the diameter used to define the circle. Hence the end of any diameter on this circle represents the value of the two dimensional stress state on any set of rotated axes relative to some original axes.

Principal Stresses

On the diameter in Figure 1-5 that lies coincident with the σ axis the normal stresses are the maximum, σ_1, and minimum, σ_2, values that will exist for any given two-dimensional stress state. These are referred to as the principal normal stresses. Their magnitudes are given by the equations that σ_1 and σ_2 are equal to the quantity $(\sigma_x + \sigma_y) / 2$ plus or minus the square root of the quantity σ_x minus σ_y over 2 quantity squared plus τ_{xy} squared.

Design for Static Mechanical Strength

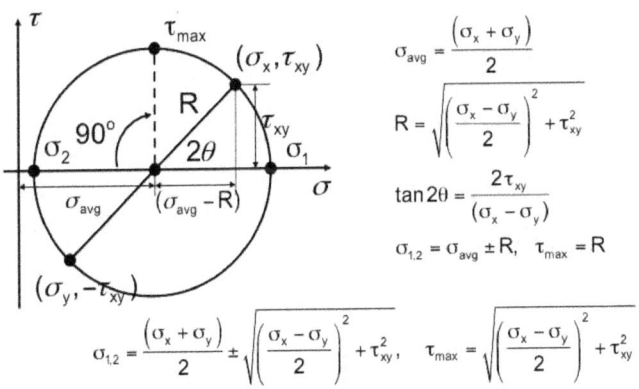

Figure 1-5 Principal Stresses

On planes of principal stresses the shear stress components are zero. The vertical diameter of the circle defines the maximum value that the shear stress can become. The magnitude of the maximum shear stress is given by τ_{max} equal to the square root of σ_x minus σ_y over 2 squared plus τ_{xy} squared. This is the second term of the principal stress equations.

The orientation of the axis system that defines the planes on which the max shear stress occurs are at 45 degrees from the axes that defines the planes on which the principal stresses occur. The values of the principal stresses associated with planes of maximum shear stress are equal to each other and are simply the average of the normal stresses, $(\sigma_x + \sigma_y)/2$, of the original stress state. The orientation of the principal stress axes with respect to the original stress state is given by θ

Design for Static Mechanical Strength

defined by the tangent of 2θ equal to $2\tau_{xy}$ divided by the quantity $\sigma_x - \sigma_y$.

Principal Stresses

$$\sigma_1, \sigma_2 = \frac{\sigma_x + \sigma_y}{2} \pm \sqrt{\left(\frac{\sigma_x + \sigma_y}{2}\right)^2 + \tau_{xy}^2}$$

Maximum Shear Stress

$$\tau_{max} = \sqrt{\left(\frac{\sigma_x + \sigma_y}{2}\right)^2 + \tau_{xy}^2}$$

$$\sigma'_x = \sigma'_y = \frac{\sigma_x + \sigma_y}{2}$$

Figure 1-6 Principal Stresses/Max Shear Stress

Special Cases

The stress state in a body in pure tension is defined by a finite σ_x stress N together with σ_y and τ_{xy} both zero. The Mohr circle representing this state of stress is tangent to the tau axis with σ_x, N, equal to the principal stress σ_1. From Figure 1-7 it is seen that the maximum shear stress is simply $\sigma_x/2$ or N/2 acting on a plane at 45 degrees from the plane on which σ_x acts.

The stress state in a body in pure torsion is represented by a Mohr circle whose center is at the origin of the sigma tau axis system since σ_x and σ_y are zero. The radius of the circle is N, the value of τ_{xy}. The principal normal stresses are given by $\sigma_1 =$ N, tension, and $\sigma_2 = -$ N, compression. Again the planes on which the maximum shear stress and principal stress act are at 45 degrees with each other.

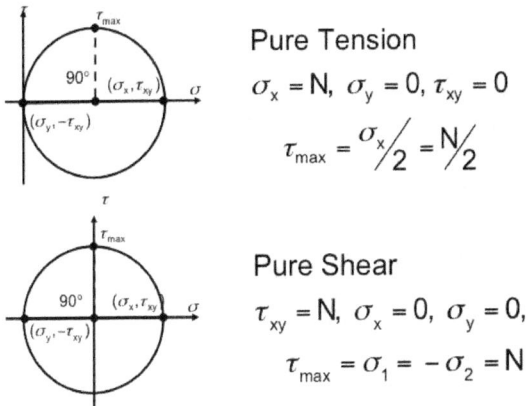

Pure Tension

$\sigma_x = N$, $\sigma_y = 0$, $\tau_{xy} = 0$

$\tau_{max} = \sigma_x/2 = N/2$

Pure Shear

$\tau_{xy} = N$, $\sigma_x = 0$, $\sigma_y = 0$,

$\tau_{max} = \sigma_1 = -\sigma_2 = N$

Figure 1-7 Special Stress states

Strain Components

The deformation of a cubical element in the xy plane due to a two dimensional state of stress is defined by elongations of the element in the x and y directions and a shearing distortion. Normal strains relative to the x and y directions are defined as elongation per unit length. For the element shown in Figure 1- 8 the normal strain in the x direction is given by ε_x equal to δ_x /L_x. In a similar fashion the normal strain in the y direction is ε_y equal to $\delta y/ L_y$.

The shear strain is defined by the angle of distortion θ_{xy} the change in the original right angle of the element. It is expressed mathematically as Υ_{xy} equal to the tangent of θ_{xy}. Since these angles

7

are very small the tangent of θ_{xy} can be replaced by the angle θ_{xy} itself.

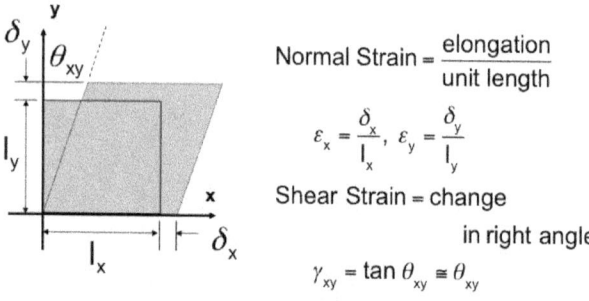

Figure 1-8 Strain Components

Principal Strains

The normal and shear strains with respect to an axis system rotated through an angle θ relative to the xy axis system are represented by the same set of transformation equations that apply to a two dimensional stress state with σ_x replaced by ε_x, σ_y replaced by ε_y and τ_{xy} replaced by $\gamma_{xy}/2$.

The Mohr circle representation for a two dimensional strain state along with the properties of principal strains and maximum shear strain are identical to that of the two dimensional stress state with the exception that the shear stress, τ, is replaced by the shear strain $\gamma/2$.

Design for Static Mechanical Strength

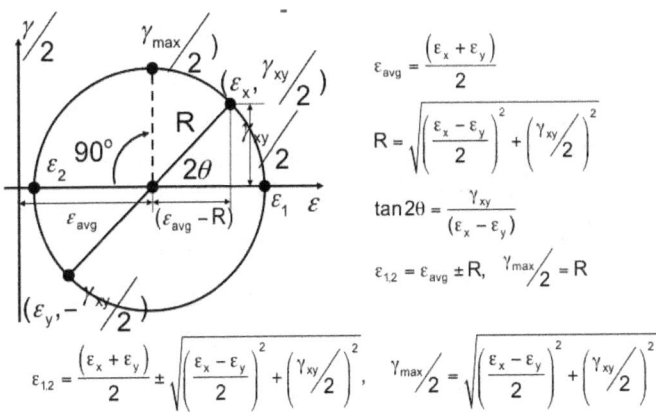

$$\varepsilon_{1,2} = \frac{(\varepsilon_x + \varepsilon_y)}{2} \pm \sqrt{\left(\frac{\varepsilon_x - \varepsilon_y}{2}\right)^2 + \left(\frac{\gamma_{xy}}{2}\right)^2}, \quad \gamma_{max}/2 = \sqrt{\left(\frac{\varepsilon_x - \varepsilon_y}{2}\right)^2 + \left(\frac{\gamma_{xy}}{2}\right)^2}$$

Figure 1-9 Mohr Strain Circle

Three D Stress/ Strain Components

A general three-dimensional stress state requires one addition normal stress and two more shear stresses. The additional normal stress is σ_z acting on the xy face (Figure 1-10) together with τ_{xz} and τ_{yz} acting on the xy face also. These three additional stresses along with those defining a two dimensional stress state make up six independent stress components.

This is similarly true with respect to the strains that define the elements distortion. The three components of a two dimensional strain must be complemented with one additional normal strain ε_z and two additional shear strain components γ_{xz} and γ yz.

Design for Static Mechanical Strength

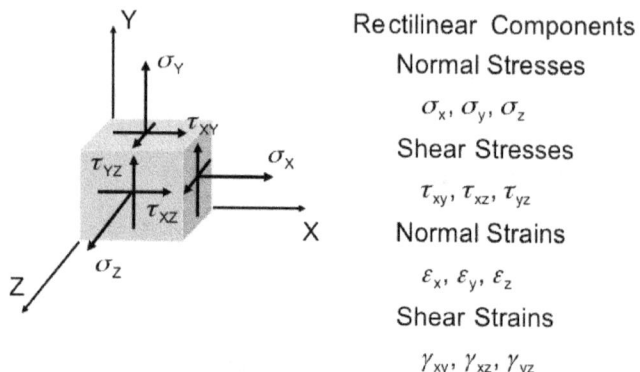

Rectilinear Components

Normal Stresses

$\sigma_x, \sigma_y, \sigma_z$

Shear Stresses

$\tau_{xy}, \tau_{xz}, \tau_{yz}$

Normal Strains

$\varepsilon_x, \varepsilon_y, \varepsilon_z$

Shear Strains

$\gamma_{xy}, \gamma_{xz}, \gamma_{yz}$

Figure 1-10 Three D Stress / Strain Components

Generalized Hooke' Law

For a homogeneous isotropic material with properties the same in all directions that behaves elastically the relationship between the strains and stresses can be represented by six linear equations (Figure 1-11). The normal strain ε_x is given by one over E, the modulus of elasticity of the material, times σ_x minus the quantity σ_y plus σ_z times v, Poison's ratio. In other words the strain is proportional to the stress in that direction but is reduced by the effect of stresses acting perpendicular to the direction of the strain.

The modulus, E, and poison's ratio, v, must be measured in the laboratory for a specific material. Recall that the modulus of elasticity, E, is the slope of the stress strain curve of a material in tension and Poisson's ratio, v, is the lateral

Design for Static Mechanical Strength

contraction perpendicular to the direction of the applied tension.

It is assumed that the shear strains are only proportional to the shear stress associated with that distortion direction. Again the shear modulus, G, must be measured for a specific material.

$$\varepsilon_x = \frac{1}{E}\left[\sigma_x - \nu\left(\sigma_y + \sigma_z\right)\right], \quad \delta_{xy} = \frac{\tau_{xy}}{G}$$

$$\varepsilon_y = \frac{1}{E}\left[\sigma_y - \nu\left(\sigma_x + \sigma_z\right)\right], \quad \delta_{xz} = \frac{\tau_{xz}}{G}$$

$$\varepsilon_z = \frac{1}{E}\left[\sigma_z - \nu\left(\sigma_y + \sigma_x\right)\right], \quad \delta_{yz} = \frac{\tau_{yz}}{G}$$

where   E = modulus of elasticity
G = shear modulus
ν = Poisson's ratio

Figure 1-11 Generalized Hooke's Law

A Special Property

Beginning with a state of pure shear as in the torsion of a shaft the associated principal stress and the resulting strains leads to a relationship between the material properties giving the shear modulus, G, equal to the modulus of elasticity, E, divided by two times the quantity 1 plus ν, the Poisson's ratio. Also since the decrease in the volume of a cubical element under hydrostatic pressure must always be greater than zero it is possible to show the Poisson's ratio must always be between zero and one half.

Substituting this restriction into the relationship between G and E results in the conclusion that the shear modulus G must always lie between one third and one half E when the material behaves elasticly.

$$G = \frac{E}{2(1+\nu)}$$

Also since volume decrease of a cubical element under hydrostatic pressure must always be ≥ 0

then $0 \leq \nu \leq 0.5$

and $\frac{E}{3} \leq G \leq \frac{E}{2}$

Figure 1-12 A Special Property

Two Special Stress State Models

Since a great number of the effects of mechanical loadings (bending, torsion and tension) can be modeled as two dimensional stress states and the mathematical manipulation of three dimensional states are difficult and complex greater emphasis is placed on two dimensional states of stress. In particular two idealized two-dimensional stress state models will be considered.

Plane Stress State

A state of plane stress is modeled by a thin plate of uniform thickness loaded by external forces in the xy plane of the plate (Figure 1-13). In this model it is assumed that σ_z is zero but ε_z exists permitting changes in thickness of the plate. This reduces the normal strain equations to functions of only two normal stress components σ_x and σ_y with the strain ε_z in the thickness dimension generated simply by a Poisson ratio effect.

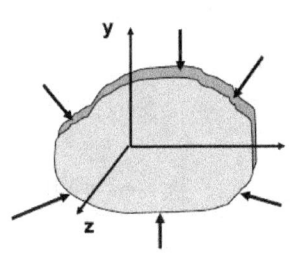

A thin plate loaded by forces applied in the xy plane

Assume $\sigma_z = 0$ hence

$$\varepsilon_x = \frac{1}{E}(\sigma_x - \upsilon\sigma_y)$$

$$\varepsilon_y = \frac{1}{E}(\sigma_y - \upsilon\sigma_x)$$

$$\varepsilon_z = -\frac{\upsilon}{E}(\sigma_x + \sigma_y)$$

Figure 1-13 Plane Stress State

Plane Strain State

A state of plane strain is modeled by a long or thick body of constant xy cross section in the z direction subjected to a uniform loading perpendicular to the z axis (Figure 1-14). This model works well for some pressure vessels, thick shrink fits and rotating disks problems.

The physical constraint of the length of the body and uniformity of load is modeled by assuming that ε_z is either zero or a constant. This results in the existence of a σ_z stress proportional to the sum of the normal stresses σ_x and σ_y. The normal strains in the x and y directions are again only functions of σ_x and σ_y with Poisson ratio effects introduced by the elimination of σ_z.

A long body loaded by constant forces perpendicular to the z axis.

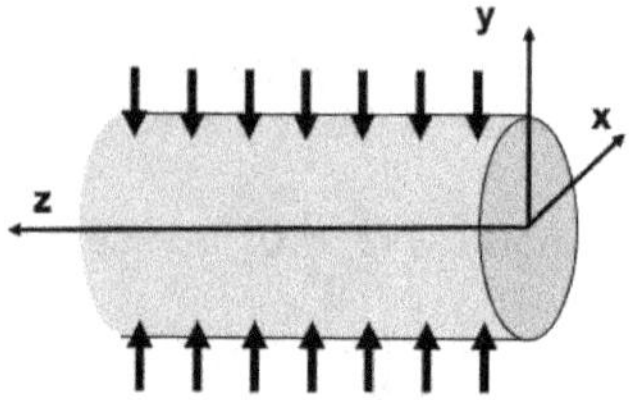

Assume $\varepsilon_z = 0$ hence

$$\sigma_z = v(\sigma_x + \sigma_y)$$

$$\varepsilon_x = \frac{1}{E}\left(\sigma_x(1-v^2) - v\sigma_y(1+v)\right)$$

$$\varepsilon_y = \frac{1}{E}\left(\sigma_y(1-v^2) - v\sigma_y(1+v)\right)$$

Figure 1 – 14 Plane Strain State

Principle Stresses – 3 D

For a given three dimensional state of stress defined by three normal and three shear stress components relative to an xyz axis system there exist some principal axis orientation x', y' and z' similar to that in a two dimensional state of stress. Relative to this orientation there will exist only normal stress σ_1, σ_2 and σ_3 on the planes perpendicular to the principal axes.

One of these will be the absolute maximum normal stress that can exist in the given original state of stress while one of the remaining principal stresses will be the absolute minimum normal stress associated with the given original state of stress. On these planes of principal stress there will be no shear stresses.

Design for Static Mechanical Strength

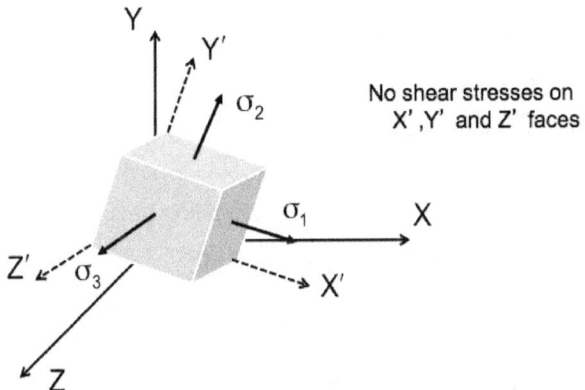

Figure 1-15 Principal Stresses – 3 D

Principal Stress Equations

The magnitudes of the three principal normal stresses are defined by a cubic equation (Figure 1-16) whose coefficients are combinations of the six components of stress that define some given original three dimensional state. The numerical solution of this equation yields three roots corresponding to σ_1, σ_2 and σ_3.

If σ_z together with τ_{xz} and τ_{yz} are set equal to zero (the classic two dimensional state of stress) the cubic equation reduces to a quadratic equation in terms of σ_x, σ_y and τ_{xy}. This equation is also given in Figure 1 – 16. The roots of this equation lead to the principal stress formulas for σ_1 and σ_2 already listed in Figure 1 – 6.

$$\sigma^3 - (\sigma_x + \sigma_y + \sigma_z)\sigma^2$$
$$+ (\sigma_x\sigma_y + \sigma_x\sigma_z + \sigma_y\sigma_z - \tau_{xy}^2 - \tau_{xz}^2 - \tau_{yz}^2)\sigma$$
$$- (\sigma_x\sigma_y\sigma_z + 2\tau_{xy}\tau_{xz}\tau_{yz} - \sigma_x\tau_{yz}^2 - \sigma_y\tau_{xz}^2 - \sigma_z\tau_{xy}^2) = 0$$

if $\sigma_z = 0$, $\tau_{xz} = 0$, $\tau_{yz} = 0$ then equation becomes

$$\sigma^2 + (\sigma_x + \sigma_y)\sigma + (\sigma_x\sigma_y - \tau_{xy}^2) = 0$$

Figure 1-16 Principal Stress Equations

Mohr's Circles – 3 D State

Assuming that σ_1, σ_2 and σ_3 are known for a given three D stress state Mohr's circles can be constructed for three pairs of these principle stresses. It will be assumed that σ_1 and sσ_3 represent the maximum and minimum values of the principal stresses.

With respect to the x'y' axes a Mohr circle can be drawn as shown in Figure 1-17 with the diameter defined by σ_1 and σ_2. Rotating the x' y' axis by 45 degrees about the z' axis produces planes perpendicular to the rotated x' y' axes on which the shear stress τ_{12}^{max} exists.

A similar construction can be generated for Mohr circles elative to the y'z' axis system involving σ_2 and σ_3 as well as a final circle relative to the x'z' axes involving the σ_1 and σ_3 stresses. It is observed that the σ_1, σ_3 circle is the largest and has associated with it the maximum shear stress τ_{13}^{max}. This is significant as this is the absolute maximum shear stress that exists for the original given state of 3 D stress.

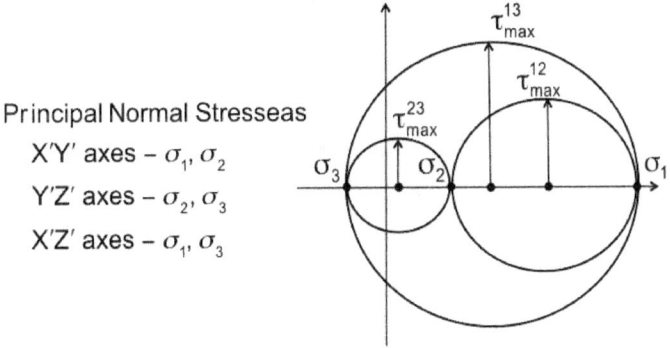

Principal Normal Stresseas
X'Y' axes – σ_1, σ_2
Y'Z' axes – σ_2, σ_3
X'Z' axes – σ_1, σ_3

Figure 1-17 Mohr Circles - 3D State

Example Problem

A steel cylindrical air tank is rated to operate at a maximum pressure of 800 psi (lbs./in.^{2}). The inside diameter of the pipe is 18 in. and the wall thickness is 1/8 in. Determine the maximum shear stress at the inner surface of the wall at some distance from the end caps. Since the ratio of the wall thickness to the radius is only slightly greater than 0.01 thin-walled pressure vessel theory stress equations can be applied.

The internal pressure produces three normal stresses on a small cubical element on the tanks inner surface. These consist of an axial (longitudinal) stress, σ_l, a circumferential (hoop) stress, σ_h and a radial stress, σ_p. There are no shear stresses acting on this element. This three dimensional stress state is shown in Figure 1 – 18.

Design for Static Mechanical Strength

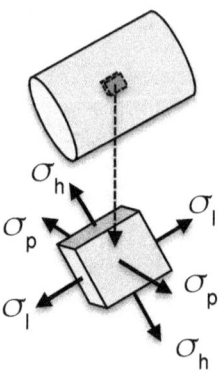

Figure 1-18 Three Dimensional Stress State

The "hoop" or circumferential normal stress is equal to the pressure times the internal radius divided by the thickness. It is calculated to be 57,600 psi (Figure 1-19). The longitudinal normal stress at some distance from the end caps will is one

Stress Calculations (thin walled theory)

Circumferencial "hoop" stress

$$\sigma_h = \frac{pr}{t} = \frac{(800)(9)}{(0.125)} = 57,600 \text{ psi}$$

Longitudinal stress

$$\sigma_l = \frac{pr}{2t} = \frac{(800)(9)}{(0.125)} = 28,800 \text{ psi}$$

Radial stress (inner wall)

$$\sigma_r = -p = -800 \text{ psi}$$

Figure 1-19 Stress Calculations

half the circumferential stress or 28,800 psi. At the inner surface the internal pressure act as a radial normal compressive stress of -800 psi.

Since there are no shear stresses on the cubical element the three stresses σ_h, σ_r and σ_l represent a three-dimensional principal stress state as shown in Figure 1 – 20

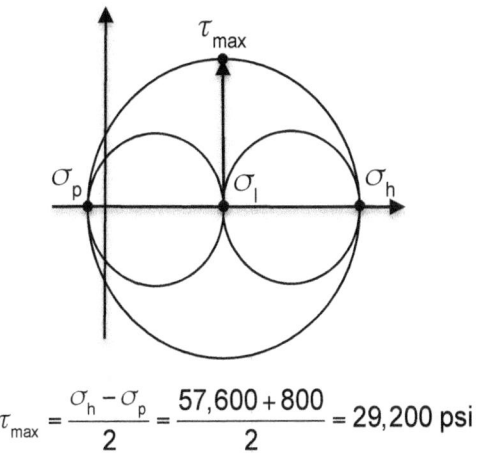

$$\tau_{max} = \frac{\sigma_h - \sigma_p}{2} = \frac{57,600 + 800}{2} = 29,200 \text{ psi}$$

Figure 1-20 Three D Principal Stress State

From the three plotted Mohr circles associated with this special state of stress it is observed that the maximum shear stress is the radius of the largest of the three circles. This maximum shear stress is calculated to be 29,200 psi.

Design for Static Mechanical Strength

Design for Static Mechanical Strength

Chapter -2 Theories of Static Failure

Chapter 2 covers static theories of failure for both ductile and brittle materials subjected to two-dimensional states of stress together with example applications

Designer's Dilemma

Ideally mechanical property data should be available from a great many tests of the material chosen for a part to insure its design provides the strength and endurance desired. It would be best if these tests could be performed on materials possessing identical alloying, processing and heat treatment to be used in the part as well as being loaded in a similar manner to the design's operational and functional requirements. This is seldom practical, possible or feasible due to conflicting issues that include safety, production quantities, testing availability as well as time and budgetary constraints.

Figure 2 -1 Designer's Dilemma

Design for Static Mechanical Strength

The designer is often faced with only being able to access published values of yield strength, tensile strength and percent elongation at failure from tensile tests. Even with this minimal information the designer must be able to estimate and predict static strength or failure under conditions that result from complex stress states. To deal with this dilemma a number of failure theories have been proposed for both ductile and brittle materials subjected to static loading.

Theories of Static Failure

The six theories presented and discussed are divided into two categories. The first is applicable to ductile materials and the second to brittle materials. The three theories for ductile materials are designated the maximum stress theory, the maximum shear stress theory and the distortion energy theory.

>**Ductile Materials**
>> Maximum Stress Theory
>> Maximum Shear Stress Theory
>> Distortion Energy Theory
>
>**Brittle Materials**
>> Mohr's Theory
>> Coulomb Mohr Theory
>> Modified Coulomb Mohr Theory

Figure 2-2 Theories of Static Failures

Design for Static Mechanical Strength

For brittle materials the three theories are Mohr's theory, the Coulomb Mohr theory and the Modified Coulomb Mohr Theory.

In presenting these theories their application is limited to two dimensional stress states. Failure for ductile materials is defined as yielding since this condition normally will impair the functionality of the design. For brittle materials failure is defined as physical facture as dimensional changes are normally small prior to failure.

Maximum Stress Theory

The criteria defining the maximum stress theory for a two-dimensional state of stress simply states that yielding of the material will occur when the applied maximum principal stress is equal to the yield stress in a simple tension test.

Four sets of principal stress conditions define yielding mathematically.
- σ_1 is equal to σ_y, (yield stress) for σ_2 between σ_y and $-\sigma_y$.
- σ_2 is equal to σ_y for σ_1 between σ_y and $-\sigma_y$.
- σ_1 is equal to $-\sigma_y$ for σ_2 between σ_y and $-\sigma_y$.
- σ_2 is equal to $-\sigma y$ for σ_1 between σy and $-\sigma y$.

These four conditions can be interpreted as the boundaries of a square plotted on a σ_1 σ_2 axis system (Figure 2-3) with the maximum values on

Design for Static Mechanical Strength

both the σ_1 and σ_2 axes equal to σ_y, the yield stress in a tension test. In this presentation the yield stress in tension and compression are assumed to be the same, a reasonable assumption for most ductile materials.

If σ_1 is always taken to be the greater of the two principal stresses then only quadrants one and four are relevant. The significance of his graphic construction is that as long as the combination of σ_1 and σ_2 represent a point that lies inside the boundary of the square in quadrants one and four yielding will not take place but as soon as one of these principal stresses lies on the boundary then yielding will take place.

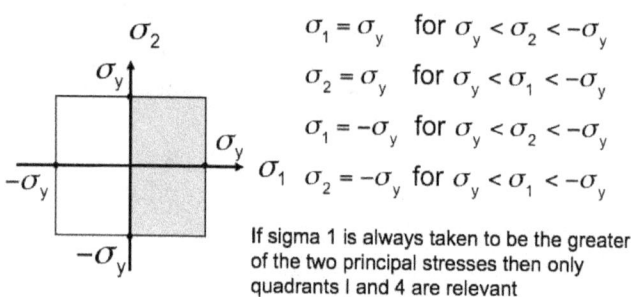

$\sigma_1 = \sigma_y$ for $\sigma_y < \sigma_2 < -\sigma_y$
$\sigma_2 = \sigma_y$ for $\sigma_y < \sigma_1 < -\sigma_y$
$\sigma_1 = -\sigma_y$ for $\sigma_y < \sigma_2 < -\sigma_y$
$\sigma_2 = -\sigma_y$ for $\sigma_y < \sigma_1 < -\sigma_y$

If sigma 1 is always taken to be the greater of the two principal stresses then only quadrants I and 4 are relevant

Figure 2-3 Maximum Stress Theory

Maximum Shear Stress Theory

The criteria that defines the maximum shear stress theory is that yielding will occur when the maximum shear stress is equal to the maximum shear stress at yielding in a tensile test. The value

Design for Static Mechanical Strength

of the maximum shear stress in a tensile test will be $\sigma_y/2$.

For a two dimensional stress state the maximum shear stress will be given by the principal stresses $(\sigma_1-\sigma_2)/2$ over two provided that σ_1 is greater than σ_2. Setting this value of τ_{max} equal to $\sigma_y/2$ gives $\sigma_1 - \sigma_2 = \sigma_y$ for the occurrence of yielding. For $\sigma 2$ greater than $\sigma 1$ the criteria for yielding becomes $\sigma_2 - \sigma_1 = \sigma_y$. These two conditions are represented graphically on the σ_1 σ_2 plot as diagonal lines through quadrants two and four (Figure 2-4).

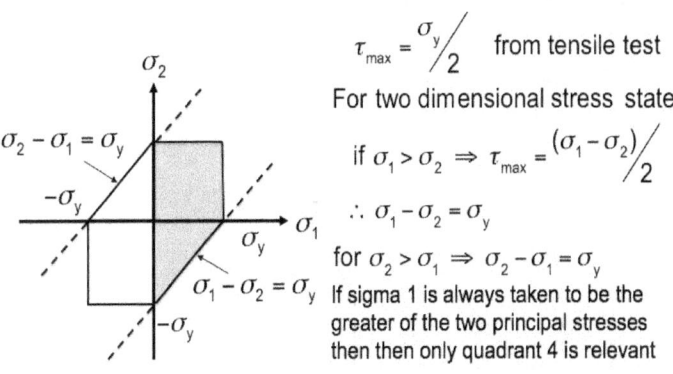

Figure 2-4 Maximum Shear Stress Theory

The theory only has significance in quadrants two and four since it predicts the occurrence of yielding at higher values of principal stress than the maximum stress theory in quadrants one and three. Again if σ_1 is always taken to be the greater of the two principal stresses then only

Design for Static Mechanical Strength

quadrants one and four are relevant. In quadrant four the maximum shear stress theory predicts yielding will occur at lower stresses than the maximum stress theory.

Yielding in Pure Shear

For a shaft in pure torsion σ_1 is equal to $-\sigma_2$ and τ_{max} is either σ_1 or $-\sigma_2$. From the maximum shear stress theory $\sigma_1 - \sigma_2 = \sigma_y$. By substituting σ_1 for $-\sigma_2$ in pure torsion results in σ_1 equal to $\sigma_y/2$.

Similarly, substituting $-\sigma_2$ for σ_1 yields σ_2 equal to $-\sigma_y/2$. Graphically pure shear is represented by a diagonal line through quadrants two and four perpendicular to the failure line for the maximum shear stress theory.

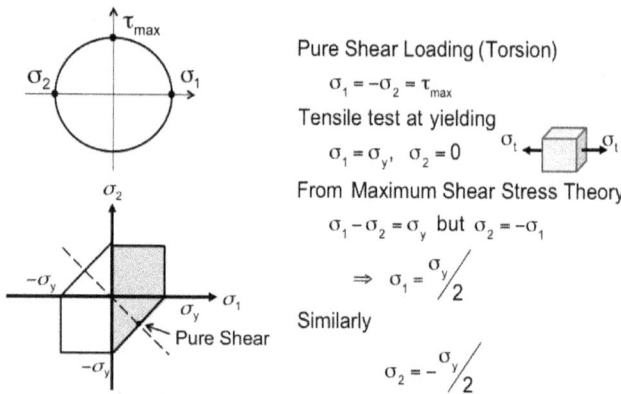

Figure 2-5 Shaft in Torsion

Design for Static Mechanical Strength

Distortion Energy Theory

The third static theory of failure states that yielding will occur when the strain energy of distortion per unit volume of material is equal to the strain energy of distortion per unit volume at yielding in a tensile test. The energy of distortion is defined as the total strain energy minus the strain energy due to volume change (Figure 2-6).

The distortion energy per unit volume will be calculated for a three dimensional stress state, set equal to the distortion energy per unit volume for a tensile test and then simplified to a two dimensional stress state.

u_s = total strain energy / unit volume
u_v = volume change strain energy / unit volume
u_d = distortioon strain energy / unit volume

$$u_d = u_s - u_v$$

Calculate u_s and u_v for principal stresses in 3D stress state and set

$$\left. u_d \right|_{\text{tensile test}} = \left. u_d \right|_{\text{3D stress state}}$$

Figure 2-6 Distortion Energy Theory

The total strain energy per unit volume is equal to the work done by the 3D stress state. This can be expressed as one half the sum of the products of the principal stresses times the principal strains.

The strains are eliminated by substituting their relationship to the stresses from Hooke's law.

Design for Static Mechanical Strength

This results in the total strain energy being represented by squares and products of the principal stresses as shown in Figure 2-7.

$$u_s = \sum_{i=1}^{3} \frac{\sigma_i \varepsilon_i}{2} = \frac{1}{2}(\sigma_1 \varepsilon_1 + \sigma_2 \varepsilon_2 + \sigma_3 \varepsilon_3)$$

but from Hooke's law (elastic material)

$$\varepsilon_1 = \frac{1}{E}(\sigma_1 - v(\sigma_2 + \sigma_3))$$

$$\varepsilon_2 = \frac{1}{E}(\sigma_2 - v(\sigma_1 + \sigma_3)), \text{ etc.}$$

so that

$$u_s = \frac{1}{2E}\left[(\sigma_1^2 + \sigma_2^2 + \sigma_3^2) - 2v(\sigma_1\sigma_2 + \sigma_1\sigma_3 + \sigma_3\sigma_2)\right]$$

Figure 2-7 Total Strain Energy

The change in volume of the unit cubical element is a consequence of the effect of a hydrostatic pressure, σ_{hp}, acting on all faces. The volume change energy is then given by substituting σ_{hp} for each principal stress into the equation for the total strain energy. This gives U_v equal to 3(1-2v)/2E times σ_{hp} squared (Figure 2-8).

Now assume that σ_{hp} can be replaced by the average of the principal stresses σ_1, σ_2, and σ_3. This results in the volume change energy as the quantity (1-2v)/6E times the sum of the principal stresses squared minus two times the products of the principal stresses. (Figure 2-8)

Design for Static Mechanical Strength

$$u_v = \frac{3(1-2v)}{2E}\left(\sigma_{hp}^2\right)$$

Assume $\sigma_{hp} = \frac{\sigma_1 + \sigma_2 + \sigma_3}{3}$ (hydrostatic pressure)

and substitute for each σ_i term in total strain energy equation to give for u_v

$$u_v = \frac{(1-2v)}{6E}\left[\left(\sigma_1^2 + \sigma_2^2 + \sigma_3^2\right) - 2\left(\sigma_1\sigma_2 + \sigma_1\sigma_3 + \sigma_3\sigma_2\right)\right]$$

Figure 2-8 Volume Change Energy

Distortion energy is defined as the total strain energy minus the volume change energy. Carrying out his subtraction results in the distortion energy equal to the quantity (1-v)/6E times the sum of the combinations of the differences of all three principal stresses squared. For a tensile test the only principal stress is σ_1 which is equal to σ_y.

$$u_d = u_s - u_v$$

$$u_d = \frac{(1-v)}{6E}\left[\left(\sigma_1 - \sigma_2\right)^2 + \left(\sigma_1 - \sigma_3\right)^2 + \left(\sigma_3 - \sigma_2\right)^2\right]$$

For tensile test $\sigma_1 = \sigma_y$, $\sigma_2 = 0$, $\sigma_3 = 0$

$$u_v = \frac{(1-2v)}{3E}\left(\sigma_y\right)^2$$

For 2D stress state

$$u_v = \frac{(1-2v)}{3E}\left(\sigma_1^2 + \sigma_2^2 - 2\sigma_1\sigma_2\right)$$

$\therefore \quad \sigma_1^2 + \sigma_2^2 - 2\sigma_1\sigma_2 = \sigma_y^2$ Distortion Energy Theory

Figure 2-9 Final Distortion Energy Equation

Design for Static Mechanical Strength

For a two dimensional stress state σ_3 is zero. Setting the distortion energy equal for these two special states leads to the final equation representing the distortion energy theory of failure for 2D stresses (Figure 2-9).

Comparison of Ductile Theories

Illustrated in Figure 2-10 are the three theories of maximum stress, maximum shear stress and distortion energy that predict when yielding will take place in a ductile material subjected to a two dimensional stress state represented in terms of principal stresses. It is interesting to note that all three theories pass through the same points on the σ_1 and sσ_2 axes.

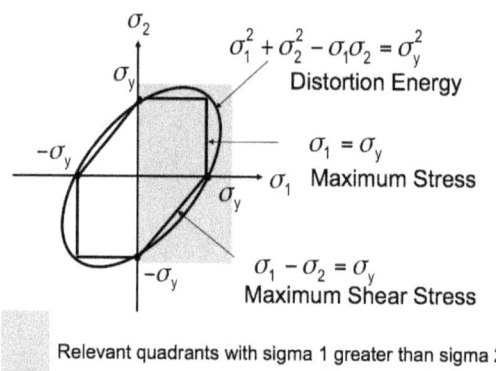

Figure 2-10 Comparison of Ductile Theories

A further observation is that all three theories do not differ by a great deal in either the first or fourth quadrants however the maximum shear stress theory is more conservative in the fourth quadrant and the maximum stress theory is more conservative in the first quadrant. Actual behavior tends to be better predicted by the distortion energy theory.

Applicable Ductile Theories

The applicability of the three theories is summarized in Figure 2-11 with the table of theories versus quadrants. Not shown in Figure 2-10 is that the maximum shear stress theory is not applicable in the first quadrant where as the maximum stress theory is not applicable in he fourth quadrant. The distortion energy theory is the least conservative of the applicable theories in both quadrants.

Quadrant	$\sigma_1 = \sigma_y$	$\sigma_1 - \sigma_2 = \sigma_y$	$\sigma_1^2 + \sigma_2^2 - \sigma_1\sigma_2 = \sigma_y^2$
First $\sigma_1 > 0$ $\sigma_2 > 0$	Yes	NA	Yes
Fourth $\sigma_1 > 0$ $\sigma_2 < 0$	NA	Yes	Yes

Figure 2-11 Summary of Ductile Failure Theories

Ductile Theory Example

Application of the theories of failure for yielding in a ductile material is demonstrated in the following example problem. A stationary solid

Design for Static Mechanical Strength

circular shaft is subjected to both a bending moment M and a twisting torque T (Figure 2-12). Using a conservative yield theory for ductile material determine a general relationship between the moment M and the torque T that will result in the onset of yielding.

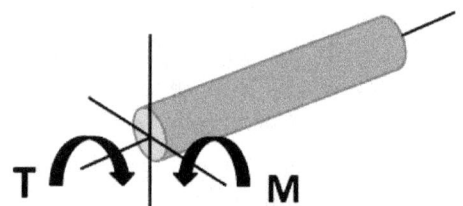

Figure 2-12 Shaft Loading

If the yield strength of the steel is 35,000 psi, the shaft diameter is 1 inch and the applied moment M is twice the torque T calculate the magnitude of the torque that will initiate yielding. The bending moment M will create a maximum normal stress σ_x of magnitude My/I where I is the moment of inertia of the cross section.

$$\sigma_x = \sigma_b = \frac{My}{I}, \quad \sigma_y = 0, \quad \tau_{xy} = \frac{Tr}{J}$$

$$\text{but } y = r \text{ and } I = \frac{J}{2}$$

$$\therefore \quad \sigma_x = \frac{2Mr}{J}, \quad \sigma_y = 0, \quad \tau_{xy} = \frac{Tr}{J}$$

therefore

$$\sigma_{1,2} = \frac{Mr}{J} \pm \sqrt{\left(\frac{Mr}{J}\right)^2 + \left(\frac{Tr}{J}\right)^2}$$

Figure 2-13 Principal Stress Calculation

Design for Static Mechanical Strength

The applied torque T will generate a maximum shear stress τ_{xy} of magnitude Tr/J where J is the polar moment of inertia of the cross section. There will be no σ_y stress. With y_{max} equal to r and I equal to J/2 then σ_x is given by 2Mr/J and max τ_{xy} is Tr/J. The principal stresses for this two dimensional stress state are given by σ_1, σ_2 equal to Mr/J plus or minus the square root of $(Mr/J)^2 + (Tr/J)^2$, (Figure 2-13).

With σ_1 positive and σ_2 negative the maximum shear stress theory is applicable as the most conservative theory for predicting ductile yielding. Substituting the principal stresses from Figure 2-13 into the shear stress yield theory equation σ_1 minus sσ_2 is equal to σ_y and simplifying the results yields the equation that $M^2 + T^2$ is equal to $\sigma_y^{2}/4$, times $(J/r)^2$, (Figure 2 – 14)

$$\sigma_1 - \sigma_2 = \sigma_y$$

$$\left\{ \frac{Mr}{J} + \sqrt{\left(\frac{Mr}{J}\right)^2 + \left(\frac{Tr}{J}\right)^2} \right\} - \left\{ \frac{Mr}{J} - \sqrt{\left(\frac{Mr}{J}\right)^2 + \left(\frac{Tr}{J}\right)^2} \right\} = \sigma_y$$

therefore

$$2\left(\frac{r}{J}\right)\sqrt{M^2 + T^2} = \sigma_y \quad \text{or}$$

$$M^2 + T^2 = \frac{\sigma_y^2}{4}\left(\frac{J}{r}\right)^2$$

Figure 2 - 14 Maximum Shear Stress Theory

Design for Static Mechanical Strength

The yield stress for the material of the shaft is 35000 psi and the loading condition is the magnitude of the bending moment is twice the magnitude of the applied torque. The polar moment of inertia is calculated to be 0.098 inches fourth.

$$\sigma_y = 35{,}000 \text{ psi, } r = 0.5 \text{ in., } M = 2T$$

$$\therefore \quad J = \frac{\pi r^4}{2} = \frac{3.14 \times 0.5^4}{2} = 0.098 \text{ in.}^4$$

$$M^2 + T^2 = 5T^2 = \frac{\sigma_y^2}{4}\left(\frac{J}{r}\right)^2$$

$$\therefore \quad T = \frac{\sigma_y}{\sqrt{20}}\left(\frac{J}{r}\right) = \frac{35}{\sqrt{20}}\left(\frac{.098}{0.5}\right) \times 10^3 = 1534 \text{ in. lb.}$$

Figure 2-15 Torque Calculation

Substituting these numerical values into the equation for the relation between M and T in the maximum shear stress theory results in the torque T to produce yielding to be 1,534 in lb (Figure 2-15).

$$\sigma_x = \sigma_b = \frac{My}{I} = \frac{4Tr}{J} = \frac{4 \times 1534 \times 0.5}{0.098} = 31{,}306 \text{ psi}$$

$$\tau_{xy} = \frac{Tr}{J} = \frac{1534 \times 0.5}{0.098} = 7{,}825 \text{ psi}$$

$$\sigma_{1,2} = \left\{ \left(31.3/2\right) \pm \sqrt{\left(31.3/2\right)^2 + 7.83} \right\} \times 10^3$$

$$\sigma_1 = 15{,}650 + 17{,}490 = 33{,}140 \text{ psi}$$

$$\sigma_2 = 15{,}650 - 17{,}490 = -1{,}847 \text{ psi}$$

Figure 2-16 Principal Stresses

Design for Static Mechanical Strength

With the applied torque T to produce yielding determined the σ_x and τ_{xy} stresses can be calculated as 31, 306 psi and 7,825 psi respectively. This permits the principal stresses σ_1 and σ_2 to be calculated as 33,140 psi and minus 1,847 psi. illustrated in Figure 2-16.

The σ_1 and σ_2 values for this problem are plotted in Figure 2-17 in the fourth quadrant of the σ_1 σ_2 axes on the boundary that defines the maximum shear stress theory.

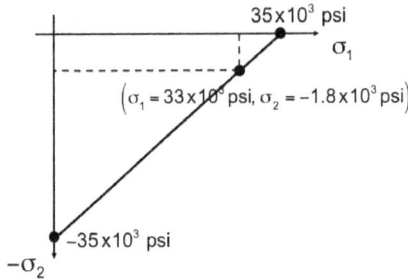

Figure 2-17 Maximum Shear Stress Theory

Mohr's Theory

Prior to presenting the theories for brittle materials some distinctions between how ductile and brittle materials are dealt with is in order. As stated earlier the yield stress in tension and compression is generally of the same magnitude in ductile materials. Also ductile materials generally under go significant elongation before facture takes place. For these reasons' yielding is typically considered to represent failure in a ductile material

Design for Static Mechanical Strength

since this state will impair the functionality of the design dimensionally.

Brittle materials more commonly are stronger in compression than they are in tension, possess lower total elongation and generally maintain dimensionality up to the point of fracture for example the failure of cast iron or glass in tension. These characteristics and behavior are expressed in the failure theories for brittle materials by basing them on total strength or fracture stress rather than yielding.

The Mohr theory of strength for brittle materials is based on test results from tensile, torsion and compression tests to failure. The theory predicts that failure will occur when the applied state of stress lies on the envelop that encloses the Mohr circle obtained from the three tests as depicted by the dotted boundaries in Figure 2-18.

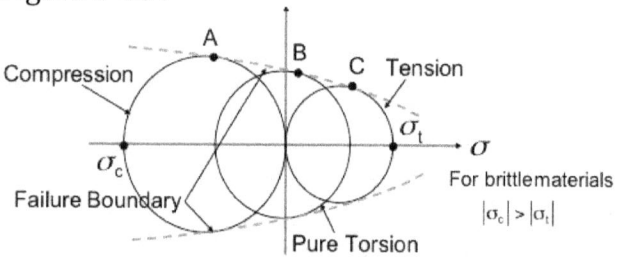

Figure 2-18 Mohr's Theory

σ_c corresponds to the compressive strength of the material and σ_t represents the stress at which tensile fracture takes place.

Coulomb Mohr Theory

The Coulomb Mohr theory is a simplified modification of the Mohr theory. It resolves the issue of analytically expressing the criteria of the Mohr theory by proposing that points A and B are tangent to the compression and tension test circles and lie on a straight line. The darker circle represents a state of stress that would represent failure.

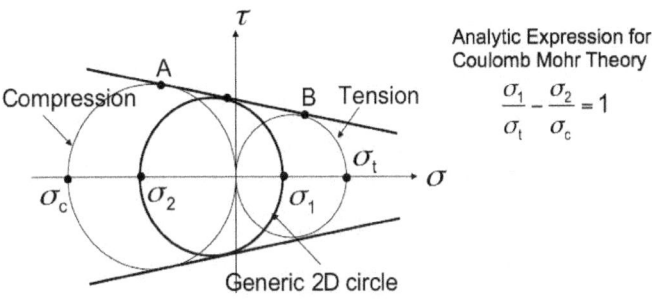

Figure 2–19 Coulomb Mohr Theory

This is expressed in a relationship involving the principal stresses σ_1 and σ_2 together with σ_c the compressive strength and σ_t the tensile strength. The equation of the boundary lines in Figure 2-19 is given by setting the ratio (σ_1/σ_t) minus the ratio (σ_2/σ_c) equal to one. A comparison of Brittle and Ductile failure theories for a two dimensional stress

Design for Static Mechanical Strength

states assuming that ductile failure is defined by tensile strength of the material is hown in Figure 2-20.

With σ_1 taken to be positive and greater than σ_2 only the first and fourth quadrants are again relevant. In the first quadrant the maximum stress theory is still dominant. In the fourth quadrant the effect of the Coulomb Mohr theory is to expand the region of the maximum shear stress theory taking into account the greater compressive strength of the brittle material.

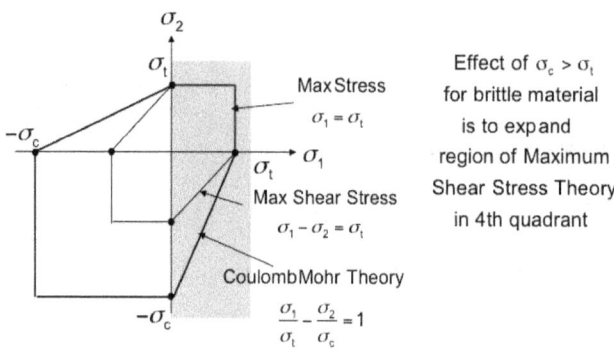

Figure 2 – 20 Ductile / Brittle Theory Comparison

Modified Coulomb Mohr Theory

The crosses in Figure 2-21 represent characteristic two dimensional stress state results in the first and fourth quadrant from the testing of brittle materials. The Coulomb Mohr theory is observed to be over conservative in the fourth quadrant. This is corrected by extending the Maximum Stress Theory down to the state of pure shear. This correction is called the Modified Coulomb Mohr Theory.

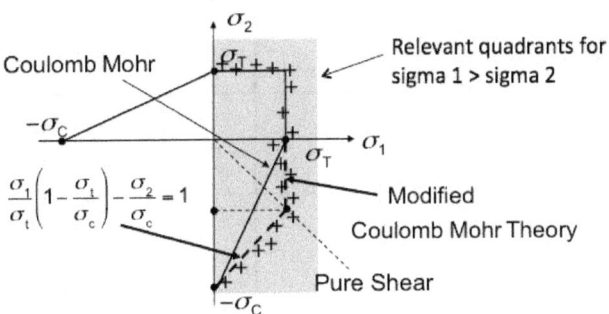

Figure 2 – 21 Modified Coulomb Mohr Theory

Analytically it is expressed for the fourth quadrant region from the state of pure shear to σ_c by the equation (σ_1/σ_t) times $(1-\sigma_t/\sigma_c)$ minus (σ_2/σ_c) equal to one as illustrated n Figure 2-21.

Applicable Brittle Theories

The applicability of the three theories for brittle materials is summarized in Figure 2-22 of theories versus quadrants. The fourth quadrant is divided into two regions. The first is defined by the condition that σ_2 is between zero and $-\sigma_t$. The remaining portion is for the condition that σ_2 is between $-\sigma_t$ and $-\sigma_c$.

The point of division between these two regions is the state of pure shear. The more

Design for Static Mechanical Strength

conservative applicable theories are indicated by "yes" underscored in the fourth quadrant while the more accurate predictions is given by the "yes" not underscored.

Quadrant	$\sigma_1 = \sigma_T$	$\dfrac{\sigma_1}{\sigma_T} - \dfrac{\sigma_2}{\sigma_C} = 1$	$\dfrac{\sigma_1}{\sigma_T}\left(1 - \dfrac{\sigma_T}{\sigma_C}\right) - \dfrac{\sigma_2}{\sigma_C} = 1$
First $\sigma_1 > 0$ $\sigma_2 > 0$	YES	NA	NA
Fourth $\sigma_1 > 0$ $0 > \sigma_2 > -\sigma_T$	YES	YES	NA
Fourth $\sigma_1 > 0$ $\sigma_2 < -\sigma_T$	NA	YES	YES

Figure 2 – 22 Applicable Theories

Brittle Theory Example

A problem is presented in Figure 2-23 to illustrate the application of the brittle material failure theories. A ceramic machine part of rectangular cross section is loaded in compression with a normal force P of 4000 lb. It is desired to determine the magnitude of torque T in inch pounds that can be applied about a longitudinal axis as shown to bring the material to the condition of potential fracture for the material properties listed. The tensile strength of the material is 20,000 psi while the compressive strength is 35,000 psi.

Design for Static Mechanical Strength

Since the compressive strength is greater than the tensile strength it is appropriate to use one of the applicable brittle theories of failure. It will be assumed that the Modified Coulomb Mohr theory is

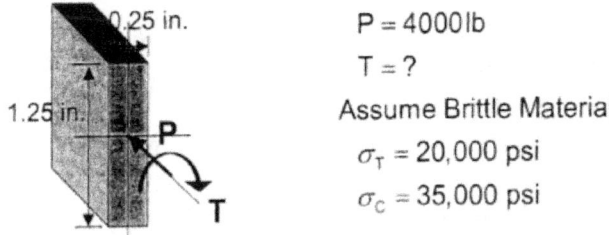

P = 4000 lb
T = ?
Assume Brittle Material
σ_T = 20,000 psi
σ_C = 35,000 psi

Figure 2 – 23   Loaded Machine Part

applicable to solve this problem. Once a solution for T is obtained it will be necessary to check if the right failure theory was used. The principal stresses must first be determined (Figure 2-24).

Calculate Stresses –

$$\sigma_x = \frac{P}{A} = \frac{-4000}{(1.25)(.25)} = -12.8 \text{ kpsi}$$

$$\tau_{xy} \cong \frac{3T}{ht^2} = \frac{3T}{1.25(.25)^2} = \frac{38T}{10^3} \text{ kpsi}$$

Principal Stresses –

$$\sigma_1 = \left\{ -6.4 + \sqrt{(6.4)^2 + \left(\frac{38T}{10^3}\right)^2} \right\}$$

$$\sigma_2 = \left\{ -6.4 - \sqrt{(6.4)^2 + \left(\frac{38T}{10^3}\right)^2} \right\}$$

Figure 2 – 24   Principal Stress Equations

41

Design for Static Mechanical Strength

Substituting the numerical values for the material properties and the given loading into the principal stress equations gives rise to a relationship that only involves the magnitude of the torque.

It is next convenient to modify the failure equation by multiplying both sides by σ_t. This makes the substitution of the principal stress expressions somewhat easier. Making the substitution for σ_1 and σ_2 leads to the numerical equation at the bottom of Figure 2–25 in which the torque T is the only unknown.

$$\frac{\sigma_1}{\sigma_T}\left(1-\frac{\sigma_T}{\sigma_c}\right)-\frac{\sigma_2}{\sigma_c}=1 \quad \text{or}$$

$$\sigma_1\left(1-\frac{\sigma_T}{\sigma_c}\right)-\sigma_2\left(\frac{\sigma_T}{\sigma_c}\right)=\sigma_T$$

$$\left\{-6.4+\sqrt{(6.4)^2+\left(\frac{38T}{10^3}\right)}\right\}\left(1-\frac{20}{35}\right)-$$

$$\left\{-6.4-\sqrt{(6.4)^2+\left(\frac{38T}{10^3}\right)}\right\}\frac{20}{35}=20$$

Figure 2 – 25   Final Torque Equation

Combining like terms in Figure 2-25 the numerical failure equation (Figure 2 –26) becomes 0.998 times the square root of $(6.4)^2 + (38\,T/10^3)^2$ equal to 19.2. Squaring both sides of this equation and solving for T gives a torque value of 473 inch pounds and a corresponding shear stress of 18,000 psi. This is the magnitude of the torsional loading that will place the principal stress state on the

Design for Static Mechanical Strength

boundary of the Modified Coulomb Mohr failure theory.

Combine like terms –

$$.998\sqrt{(6.4)^2 + \left(\frac{38T}{10^3}\right)^2} = 19.2$$

Square both sides –

$$41 + \left(\frac{38T}{10^3}\right) = 366 \Rightarrow T = \frac{18}{38} \times 10^3 = 473 \text{ inlbs} \Rightarrow \tau \cong 18 \text{ kpsi}$$

Verify choice of theory –

$$\sigma_1 = \left(-6.4 + \sqrt{(6.4)^2 + 18^2}\right)$$
$$= (-6.4 + 19.1)10^3$$
$$\sigma_1 = 12,700 \text{ psi}$$
$$\sigma_2 = (-6.4 - 19.1)10^3$$
$$\sigma_2 = -25,500 \text{ psi}$$

Figure 2 – 26   Stress Calculations

To determine if the correct theory was used the magnitude of the principal stresses are now calculated. This gives σ_1 equal to 12,700 psi and σ_2 equal to minus 25,500 psi. This combination places the principal stress coordinates in the lower portion of the fourth quadrant where the Modified Coulomb Mohr theory is applicable.

Illustrated in Figure 2-27 is an approximate graphical representation of where the coordinate point of the principal stresses falls in the fourth quadrant on the boundary of the Modified Coulomb Mohr portion of the failure diagram.

Design for Static Mechanical Strength

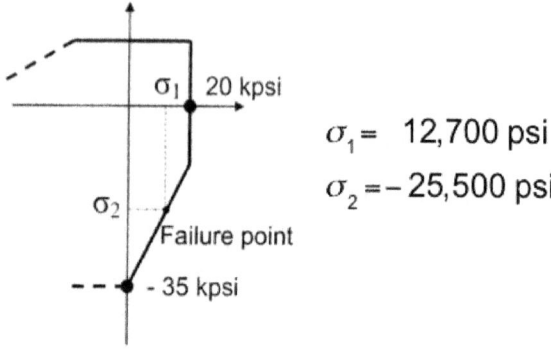

Figure 2 – 27 Modified Coulomb Mohr Diagram

Design for Static Mechanical Strength

Chapter -3 Factors of Safety

Chapter 3 deals with introducing the concept of a factor of safety in mechanical design for the purpose of minimizing the risk of potential failure from incomplete or questionable knowledge about those factors that affect the adequacy of the design. The issues addressed include: the effect of variability in loading, part geometry and material properties, inclusion of a factor of safety in classic theories of failure and statistical failure probability determination in large product populations.

Definition

The simplest definition of a factor of safety is the ratio of the strength of the material, normally expressed as the yield or tensile strength, to the maximum working stress carried by the design as a consequence of external loading and geometry.

$$N = \frac{\sigma_m}{\sigma_w}$$

N = factor of safety
σ_w = working stress psi
σ_m = material strength psi
    (yield or tensile strength)

Figure 3-1 Factor of Safety Definition

Design for Static Mechanical Strength

For example if the factor of safety for a specific design is two the allowable external loading will produce an internal working stress of magnitude that is only one half the magnitude of the strength of the material specified as predicting that failure will occur.

Variation Scenario

To illustrate numerically the role of the factor of safety where there is incomplete or questionable knowledge in a design, consider the bending of a beam with a possible variation in the loading, the dimensions of the beam cross-section and the properties of the material of construction.

For example consider that an applied critical bending moment of 10,000 in. lbs. can vary by plus or minus 500 in. lbs. The nominal height of the beam's rectangular cross section is two inches but can vary by plus or minus one eight of an inch. Similarly the width of the cross section is one inch plus or minus one sixteenth of an inch.

Loading:
  M = 10,000 ± 500 in. lbs.
Cross section dimensions:
  h = 2 ± 1/8 in.
  w = 1 ± 1/16 in.
Yield stress:
  σ_y = 15,000 ± 1,500 lb./in.^{2}

Figure 3-2 Example Parameters

Design for Static Mechanical Strength

As illustrated in Figure 3-3 the nominal working bending stress is calculated to be 14,900 psi based on nominal loading and cross section dimension values. However, the maximum working bending stress based on maximum loading and minimum cross section dimensions is 19,100 psi. This is almost 30 % higher than the nominal working stress.

Nominal bending Stress:

$$\sigma = \frac{Mc}{I}, \quad I = \frac{bh^3}{12} = \frac{1 \times 2^3}{12} = 0.667 \text{ in.}^4$$

$$\sigma_w^{nom} = \frac{10,000 \times 1}{0.667} = 14,990 \text{ psi}$$

Maximum bending Stress:

$$\sigma_w^{max} = \frac{M_{max} c_{min}}{I_{min}} = \frac{10,500 \times .9375}{.515} = 19,110 \text{ psi}$$

$$\sigma_w^{max} / \sigma_w^{nom} = \frac{19,110}{14,990} = 1.27 \quad \Rightarrow \quad \sigma_w^{max} = 1.27 \, \sigma_w^{nom}$$

Figure 3-3 Nominal/Maximum Working Stress

Now consider the effect of variability in the value of the yield stress of the material, i.e. σ_y is 15,000 psi plus or minus 1500 psi. The minimum yield strength would be 13,500 psi or nine tenths of the nominal yield stress. For a safe design based on the maximum stress theory the maximum working stress must be equal to or less than the minimum yield stress.

Design for Static Mechanical Strength

For a safe design (Max Stress Theory)

$\sigma_w^{max} \leq \sigma_y^{min}$ with $\sigma_w^{max} = 1.27\sigma_w^{nom}$ & $\sigma_y^{min} = 0.9\sigma_y^{nom}$

Introduce Factor of Safety

$1.27\sigma_w^{nom} \leq 0.9\sigma_y^{nom}$ but $N = \dfrac{\sigma_y^{nom}}{\sigma_w^{nom}}$

$\therefore \quad N \geq \dfrac{1.27}{0.9} = 1.41$

Figure 3-4 Resultant Factor of safety

These stress values are now replaced by their representation in terms of nominal stresses taking into the account the uncertainty of the loading, dimensions and material properties. Introducing the concept of a factor of safety into the requirement for a safe design results in a calculated value of N equal to 1.41, (Figure 3-4). For the design to be based simply on nominal values of loading, dimensions and material properties a factor of safety of 1.41 would need to be included to insure that the worst case introduced by the uncertainties was covered.

Minimum Factor of Safety

Assuming that load, geometry and material properties variations are known or can be estimated a generic process for establishing the minimum required factor of safety can be formulated. From nominal load and geometric values together with standard models for predicting stress the nominal working stress is calculated. In a similar fashion

Design for Static Mechanical Strength

values from estimated or known variations in the nominal design values can be used to calculate a maximum working stress.

The ratio of the maximum working stress to the nominal working stress is designated as C_{load} or C_L indicating how much greater the maximum working stress is than the nominal working stress (Figure 3-5).

From nominal load and geometry values
calculate nominal working stress – σ_w^{nom}
From extreme values
calculate maximum working stress – σ_w^{max}
Determine ratio

$$\sigma_w^{max} / \sigma_w^{nom} = C_{load} = C_L$$

Figure 3-5 Working Stress Ratio

Material property variations are handled in a similar fashion. The ratio of the minimum yield stress to the nominal yield stress is designated $C_{capability}$ or C_C. For a safe design the maximum working stress is set equal to or less than the minimum yield stress.

Substituting for these stress values their representation in terms of nominal stress values and correction constants C the factor of safety can be expressed as the ratio of C_L over C_C. This is the minimum factor of safety that will account for

Design for Static Mechanical Strength

variations in the loading, dimensions, stress model and material properties.

From variation in strength properties (yield)

calculate $\dfrac{\sigma_y^{min}}{\sigma_y^{nom}} = C_{capability} = C_C$

For a safe design

$\sigma_w^{max} \leq \sigma_y^{min}$ or $C_L \sigma_w^{nom} \leq C_C \sigma_y^{nom}$

but $\left(\dfrac{\sigma_y}{\sigma_w}\right)^{nom} \geq \dfrac{C_L}{C_C} = N$ Factor of safety

Figure 3-6 Minimum Factor of Safety

Apply to Max Shear Stress Theory

The presentation thus far has been restricted to a single working stress. Now consider the application of the generic minimum factor of safety concept to the theories of failure covered in Chapter 2. This is demonstrated by its application to the maximum shear stress theory of failure.

It is assumed that the correction for stress calculations C_L can be applied equally to the principal stresses in a two- dimensional stress state. $(\sigma_1 - \sigma_2)_{max}$ is set equal to $(\sigma_1 - \sigma_2)_{nom}$ times C_L. Also $(\sigma_y)_{min}$ is expressed as $(\sigma_y)_{nom}$ times C_c For a safe design $(\sigma_1 - \sigma_2)_{max}/(\sigma_y)_{min}$ must be equal to or less than one.

Substituting into this relationship the stresses expressed in nominal terms with the correction

Design for Static Mechanical Strength

constants C leads to the maximum shear stress theory now expressed as the quantity $(\sigma_1 - \sigma_2)/(\sigma_y)$ equal to one over the factor of safety N where all the stresses are nominal. See Figure 3-7 for this development.

Max Shear Stress Theory

$$\sigma_1 - \sigma_2 \leq \sigma_y$$

with load and capability variations

$$(\sigma_1 - \sigma_2)^{max} = (\sigma_1 - \sigma_2)c_L, \quad (\sigma_y)^{min} = (\sigma_y)c_C$$

so that

$$\frac{(\sigma_1 - \sigma_2)^{max}}{(\sigma_y)^{min}} = \frac{(\sigma_1 - \sigma_2)c_L}{(\sigma_y)c_C} \leq 1 \Rightarrow$$

$$\frac{\sigma_1 - \sigma_2}{\sigma_y} \leq \frac{c_C}{c_L} = \frac{1}{N}$$

Figure 3-7 Max Shear Stress Theory with FS

Modified Theories of Failure

In a manner similar to that employed to introduce a factor of safety into the maximum shears tress theory of failure all of the theories of failures presented in Chapter 2 can be in put in the form of having the material properties on the left side of the equation with the magnitude of these stress ratios equal to or less than one over the factor of safety. These modified formulations are presented in Figure 3–8.

In these representations it is assumed that that C_L is the same for both principal stresses and C_c is

Design for Static Mechanical Strength

the same for the tensile and compressive yield stress in ductile materials as well as the tensile and compressive strengths in brittle materials. It is recognized that this is an approximation but allows for simpler representation of the effect of including a factor of safety.

Theory	Expression
Normal Stress	$\dfrac{\sigma_1}{\sigma_y} \leq \dfrac{1}{N}$
Maximum Shear Stress	$\dfrac{\sigma_1 - \sigma_2}{\sigma_y} \leq \dfrac{1}{N}$
Distortion Energy	$\dfrac{\left(\sigma_1^2 + \sigma_2^2 - \sigma_1 \sigma_2\right)^{1/2}}{\sigma_y} \leq \dfrac{1}{N}$
Coulomb Mohr Theory	$\dfrac{\sigma_1}{\sigma_T} - \dfrac{\sigma_2}{\sigma_C} \leq \dfrac{1}{N}$
Modified Coulomb Mohr Theory	$\dfrac{\sigma_1}{\sigma_T}\left(1 - \dfrac{\sigma_2}{\sigma_C}\right) - \dfrac{\sigma_2}{\sigma_C} \leq \dfrac{1}{N}$

Figure 3-7 Modified Failure Theories with FS

Effect of Factor of Safety

The effect of introducing a factor of safety on the boundaries of the maximum stress and maximum shear stress theories for a ductile material is illustrated in Figure 3-8. As developed in Chapter 2 all of the theories of failure possess a factor of safety of one. With a factor of safety of two the maximum values of allowable stress on the σ_1 and σ_2 axes are $\sigma_y/2$ whereas on the minus σ_2 axis the maximum value is minus $\sigma_y/2$. The area of acceptable design has been significantly reduced.

Design for Static Mechanical Strength

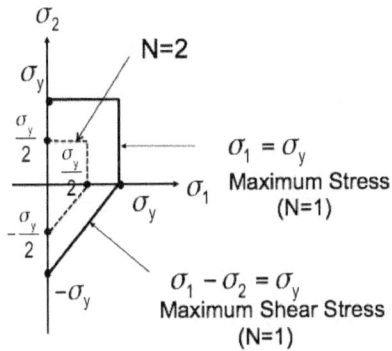

Figure 3-8 Effect of Factor of Safety

Statistical Factor of Safety Treatment

When parts are produced in very large numbers actual load and load capabilities can have probability distributions as depicted in Figure 3-9.

Probability density is plotted vertically versus actual load and load capability distributions horizontally. The nominal factor of safety N is defined in terms of the ratio of the mean values of these two distributions. These are defined for the actual load as L bar and for the load capability as L_c bar. The load range where these two distributions overlap represents possible failure occurrences. That is, the actual load on these parts is greater than the load capability of the design.

Design for Static Mechanical Strength

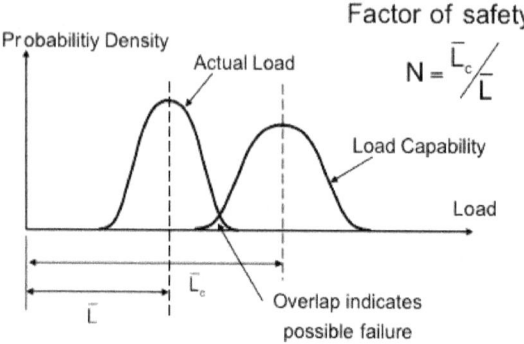

Figure 3-9 Load and Load Capability Distributions

If the load probability distribution is subtracted from the load capability distribution the overlap area in Figure 3 – 10, shown shaded to the left of the origin under the combined probability curve, represents the percentage of the total number of parts that will fail.

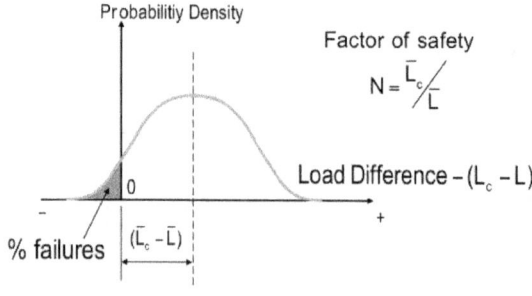

Figure 3-10 Percent Failure Representation

Design for Static Mechanical Strength

This shaded area can be mathematically related to the nominal factor of safety N defined by L_c bar over L bar by assuming the probability difference distribution is Gaussian.

Apply Gaussian Distribution

Assuming the L_c minus L distribution to be approximately Gaussian an expression for the factor of safety can be written as shown in Figure 3–11. The parameter t_f is the upper limit of the integral of the shaded failure area portion of the Gaussian approximation. In this representation t_f is related to a specific decimal percentage of failures to be described in detail. D_L is the standard deviation of the applied load approximated by ΔL, the variation in the nominal load, divided by three.

$$\frac{\bar{L_c}}{\bar{L}} = N = 1 + \frac{t_f \sqrt{(D_{L_c})^2 + (D_L)^2}}{\bar{L}} = FS$$

where

t_f = upper limit of $\int$ representing failure area (decimal percentage of failures)

D_L = standard deviation of applied load $\cong \Delta L/3$

D_{L_c} = standard deviation of load capability $\cong \Delta L_c/3$

Figure 3-11 Statistical Factor of Safety Expression

In a similar fashion D_{LC} is the standard deviation of the nominal load capability, approximated by ΔL_C divided by three.

Design for Static Mechanical Strength

Example Values of t_f

A portion of the Gaussian probability function table is presented in Figure 3–12. This permits the determination of the magnitude of t_f for a specified failure rate. Assume that a rate of failure not to exceed

t_f	0	0.02	0.04	0.06	0.08
2.0	0.02280	0.02170	0.00206	0.01970	0.01880
2.2	0.01390	0.01320	0.01250	0.01190	0.01130
2.4	0.00820	0.00776	0.00415	0.00391	0.00366
2.6	0.00466	0.00440	0.00415	0.00391	0.00368
2.8	0.00250	0.00240	0.00220	0.00212	0.00190
3.0	0.00135	0.00069	0.00034	0.00016	0.000070

Figure 3-12 Portion of Gaussian Probability Table

125 parts per 10,000 or 1.25 parts per 100 is desired. This corresponds to a percentage failure rate of 1.25 percent corresponding to a decimal percentage failure rate of 0.0125.

This value is located in the matrix of the table in the shaded block. This location is at the intersection of the 2.2 block in the vertical left column and the 0.04 block in the horizontal top row. The value of t_f for this failure rate is the sum of these two numbers, 2.24. This table covers failure rates from roughly 230 parts per 10,000 down to just under 1 part per 10,000.

Design for Static Mechanical Strength

Sample Problem
Determining a statistical factor of safety is illustrated with the following sample problem. Assume that a bar in tension is to carry a load of 1000 lbs. with a possible variation of plus or minus 25 percent. Its maximum load capability must be at least 1250 lbs. But the load capability can only be known to within plus or minus 10 percent variation. Assume further that enough parts are to be produced such that the variations are distributed normally or Gaussian.

Calculate the necessary factor of safety to be applied to the mean values of load and load capability if the failure rate is not to exceed 0.05 % (5 parts per 10,000) or a decimal percentage failure rate of .0005. The applied load is 1000lbs. and ΔL is 250 lbs. Assume that the nominal load capability is 1300 lbs, 50 lbs. over the maximum required load capability. This makes the Δload capability 130 lbs.

Applied load
$\bar{L}$ = 1000 lbs., ΔL = 250 lbs.
Load capability(assume)
$\bar{L}_c$ = 1300 lbs., ΔL_c = 130 lbs

Figure 3-13 Example Problem Parameters

Problem solution
The standard deviation of the load capability is 130 divided by 3 or 43.3 and the standard deviation of the applied load is 250 divided by three

Design for Static Mechanical Strength

or 83.3. For a failure rate of 5 parts per ten thousand corresponding to a decimal percentage rate of 0.0005 the value of tf is

$$\frac{\bar{L}_c}{L} = N = 1 + \frac{t_f \sqrt{(D_{L_c})^2 + (D_L)^2}}{L} = FS$$

$$D_{L_c} = \frac{\Delta L_c}{3} = \frac{130}{3} = 43.3 \qquad D_L = \frac{\Delta L}{3} = \frac{250}{3} = 83.3$$

For decimal percentage failure rate of 0.0005 the value of t_f from table on slide 14 is interpolated to be 3.03

Then

$$N = 1 + \frac{3.03\sqrt{(43.3)^2 + (83.3)^2}}{1000} = 1.28 = FS$$

Figure 3-14 Factor of Safety Calculation

interpolated from the table in Figure 3-12 to be 3.03. Substituting these values into the equation for the factor of safety from page 13 gives a value N equal to 1.28 in Figure 3 - 14.

Additional Approximations

For failure rates between one to fifteen parts per thousand the statistical factor of safety equation can be approximated by the equation in Figure 3-15 with t_f replaced by 1.29 and a term added to the denomination which is the decimal percentage failure raised to the 0.128 power. This will give a result that is within a few percent of the more accurate equation. For zero failures the FS is given by the relation 1 plus ΔL over L bar divided by one minus ΔL_c over L_c bar.

Design for Static Mechanical Strength

For failure percentages in the range (0.1 to 1.5%), (1 to 15 parts per 1000) a good approximation is

$$N = 1 + \frac{1.29\sqrt{\left(D_{L_c}\right)^2 + \left(D_L\right)^2}}{\left(F_d\right)^{0.128} \bar{L}} \quad \text{(to within} \approx 2\%\text{)}$$

where F_D = decimal % failure ($.0001 \le F_D \le .015$)

For zero failures:

$$N \ge \frac{1 + \left(\Delta L / \bar{L}\right)}{1 - \left(\Delta L_c / \bar{L}_c\right)}$$

Figure 3-15 Factor of Safety Approximations

If the approximation in Figure 3-15 is applied to the sample problem the factor of safety is calculate to be 1.32 for a failure rate of 5 parts per 10,000. This is within less than three percent of the previous value determined. For a zero failure rate the factor of safety increases to 1.39 or 8.6 percent higher than the lower factor of safety.

$$N = 1 + \frac{1.29\sqrt{\left(D_{L_c}\right)^2 + \left(D_L\right)^2}}{\left(F_d\right)^{0.128} \bar{L}} = 1 + \frac{1.29\sqrt{(43.3)^2 + (83.3)^2}}{(0.0005)^{0.128}(1000)}$$

N = 1.32 (less than 3% different from previous result)

For zero failures:

$$N \ge \frac{1 + \left(\Delta L / \bar{L}\right)}{1 - \left(\Delta L_c / \bar{L}_c\right)} = \frac{1 + .25}{1 - 0.9} = 1.39$$

Figure 3-16 Approximate FS Calculation

Design for Static Mechanical Strength

Design for Static Mechanical Strength

Chapter -4 Stress Concentration Factors

Chapter 4 deals with the increase in nominal stress values at sudden changes in part geometry. This is accounted for by multiplying the nominal stress by a correction, a stress concentration factor.

Occurrence of Phenomena

A circular rod of cross section A_1 loaded with a tension force P generates an internal nominal stress σ_1 equal to P/A_1. If a bar of area A_2, greater than A_1, is subjected to the same tension force P the nominal stress σ_2 will be P/A_2. with $\sigma_2 < \sigma_1$.

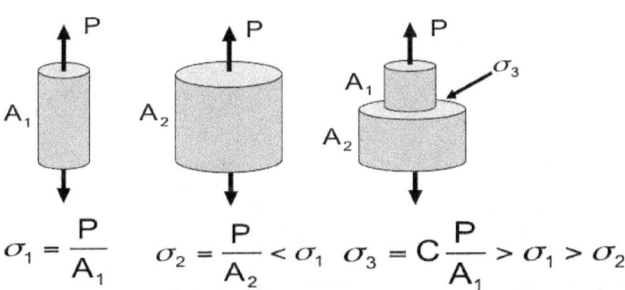

$\sigma_1 = \dfrac{P}{A_1}$ $\sigma_2 = \dfrac{P}{A_2} < \sigma_1$ $\sigma_3 = C\dfrac{P}{A_1} > \sigma_1 > \sigma_2$

Figure 4-1 Stress Concentration Occurrence

If the bar transitions rapidly from A_2 to A_1 as shown on the right of Figure 4–1 the stress at the transition sσ_3 will be greater than σ_1 or σ_2.

Design for Static Mechanical Strength

Its value is classically represented as the greater of σ_1 and σ_2 multiplied by a correction factor C. This correction is designated the stress concentration factor. It magnitude is dependent on the severity of the geometric transition and the type of load applied.

Flow Visualization–

This physical increase in stress at a point of transition in geometry can be visualized by considering that stress can be modeled like laminar fluid flow in a pipe and picturing how the fluid stream lines behave as they pass the point of transition. In Figure 4 -2 the stream-lines on the left are crowded together and undergo dramatic direction changes as the flow proceeds from left to right. The more severe these streamline changes are to one another the higher the stress in that vicinity.

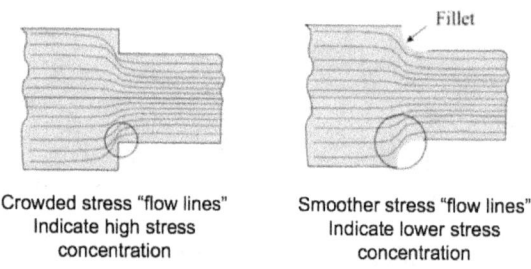

Crowded stress "flow lines" Indicate high stress concentration

Smoother stress "flow lines" Indicate lower stress concentration

Figure 4-2 Flow Visualization

Design for Static Mechanical Strength

On the right the geometric transition has been relieved by the introduction of a fillet. In this case the fluid streamlines are less crowded and smoother as they pass the geometric change in the section. The increase in stress in that vicinity is less than the transition on the left but it still represents a stress increase over the higher nominal stress in the bar.

Symbolic Representation

The magnitude by which the stress increases over the nominal stress at a sudden geometric change is defined as the stress concentration factor K_t for normal stresses and K_s for shear stress. This is expressed analytically as K_t equal to σ_{max} over $\sigma_1^{nominal}$ and K_s equal to τ_{max} over τ_1^{nomina}.

$$\sigma_{max} = K_t \, \sigma_1^{nominal}$$
$$\text{where } \sigma_1^{nominal} > \sigma_2^{nominal}$$

$$\tau_{max} = K_s \, \tau_1^{nominal}$$
$$\text{where } \tau_1^{nominal} > \tau_2^{nominal}$$

Figure 4-3 Symbolic Representation

These nominal values are the greater of the two stresses that exist across the geometric transition. The factors K_t and K_s are functions of the shape of the geometric change and the type of loading being applied.

A Theoretical Solution

The theory of elasticity can be used to calculate the magnitude of the stress concentration for a limited number of changes in geometry. The exact solution for the stress concentration factor K_t for a plate under tension that contains a generic elliptical hole defined by the two half axes a and b is illustrated in Figure 4-4.

As long as the width of the plate c can be considered large compared to the dimension b the theoretical value of K_t is given by $(1 + 2b/a)$. For the special case of a equal to b which defines a circular hole the value of K_t is three. If a is equal to b over 2 producing an ellipse the value K_t increases to 5. For the limiting case of a approaching zero which would appear as a horizontal crack K_t approaches infinity.

From "Theory of Elasticity"

$K_t = 1 + \dfrac{2b}{a}$ $c >>> b$

If a = b (circle)

$K_t = 3$

If a = b/2

$K_t = 5$

If a $\Rightarrow$ 0 (crack)

$K_t \Rightarrow \infty$

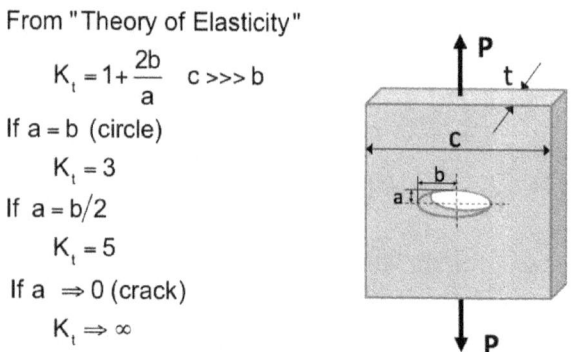

Figure 4-4 A Theoretical Solution

K_t variation with a/b

Illustrated in Figure 4-5 is the variation of Kt with the ratio of a over b. As a over b approaches zero physically representing a horizontal crack the value of K_t will approach infinity.

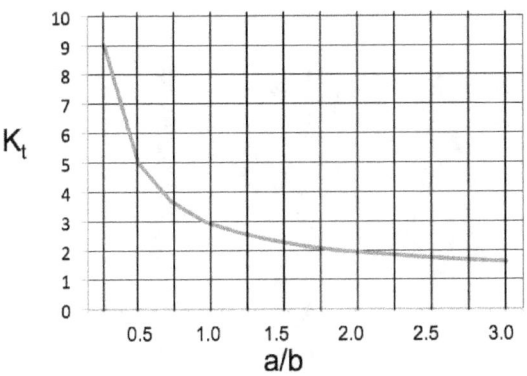

Figure 4-5 Variation of Kt with a/b

As a over b becomes very large approaching infinity the hole becomes a vertical crack and K_t approaches one indicating there is no stress concentration effect. Both of these limiting condition seem very reasonable from simple physical considerations.

Modeling Techniques

Stress concentration factor values can be determined experimentally using a technique called photo-elasticity. Polarized light passed through a loaded birefringent transparent model of the geometric shape change creates optical patterns

Design for Static Mechanical Strength

directly related to the state of stresses at that point. Maximum stress values can be determined by analyzing the resulting colored fringe patterns. Illustrated in Figure 4-6 are photoelastic patterns in three models of a plate in tension with increasing degrees of geometric change for the same minimum cross section.

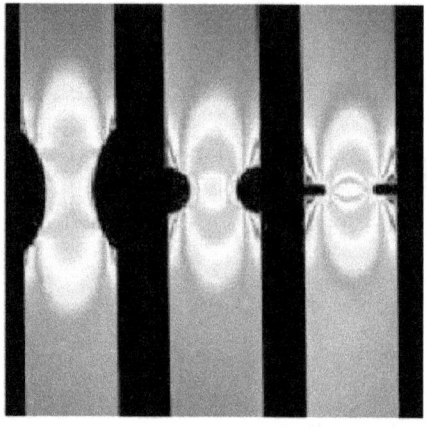

Figure 4-6 Photoelastic Specimens

The change in the pattern is more severe and there are more fringes generated in going from left to right. This indicates a higher level of stress concentration whose numerical value can be determined using photo elastic theory. This is how stress concentration factors were determined prior to the advent of digital computers and finite element software.

Design for Static Mechanical Strength

Today, modern CAD software coupled with finite element analysis capability permits three-dimensional modeling of a loaded physical part and the creation of colored stress fields that indicate where stresses are maximum and their magnitude relative to some nominal base value. Illustrated in Figure 4-7 is a stepped circular bar under axial tension. The dark gray and light gray fields indicate nominal uniform stresses with light gray being the higher stress value. At the transition point the intermediate gray band represents a significantly higher value of stress before tapering off to the lower stress n the larger section.

Figure 4-7 Stepped Shaft FEA Results

Published Information

The quintessential source of published stress concentration information is the third edition of Peterson's Stress Concentration Factors edited by Pilkey and Pilkey. This source of information is presented in chart form for a variety of geometric transitions like steps, shoulders, grooves, holes, etc.

Design for Static Mechanical Strength

in plates and shafts subjected to static tension, bending and torsion in terms of geometric parameters that define the transition. Examples of these charts are presented in the next three figures.

Stress Concentration Factors (shoulder fillet – axial load)

Figure 4-8 provides K_t values for a stepped shaft with a fillet shoulder under axial tension. The geometry is characterized by the ratio of the diameters

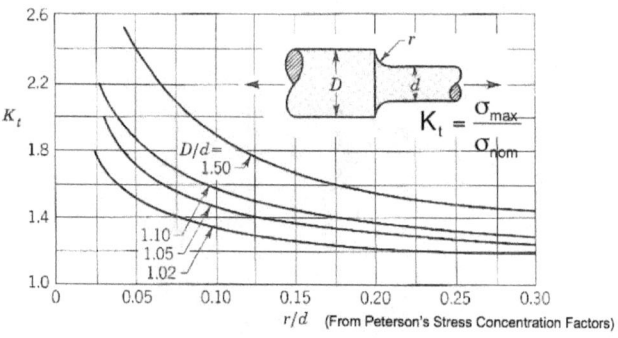

Figure 4-8 K_t for Stepped Shaft in Tension

of the two section, D over d and the ratio of the fillet radius r to the smaller diameter d. For a constant value of r/d, K_t is seen to increases as the shaft diameter difference increases. For a constant value of D/d the K_t value increases at an increasing rate as the ratio of the fillet radius to the small diameter decreases.

Design for Static Mechanical Strength

Stress Concentration Factors (shoulder fillet – bending load)

The K_t factor values in Figure 4-9 are for a similar geometric shoulder fillet transition but now for a bending load M. The graphs are again characterized by the ratio of D/d a function of the fillet radius to the smaller diameter. It is observed that for the same value of D/d of 1.5 and fillet radius to smaller diameter of 0.05 the magnitude of K_t for bending is about 2.1 compared to 2.4 for the shaft in tension demonstrating that the loading as well as the geometry affects the amount by which the local stress in increased.

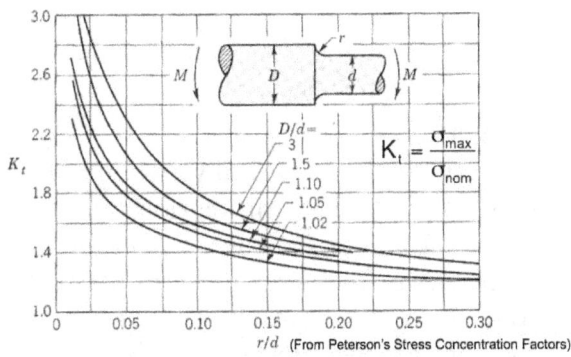

Figure 4-9 K_t for Stepped Shaft in Bending

Stress Concentration Factors (shoulder fillet – torsion loading)

Figure 4-10 presents the variation of K_s for a stepped shaft again characterized by the diameter

Design for Static Mechanical Strength

ratio and fillet radius to smaller diameter for an applied torsion load. It is observed here that for D/d of 1.5 and r/d of 0.05 the Ks factor is about 1.7 indicating that this transitional change has a smaller effect on shear stress due to torsion compared to the normal stresses generated by tension or bending.

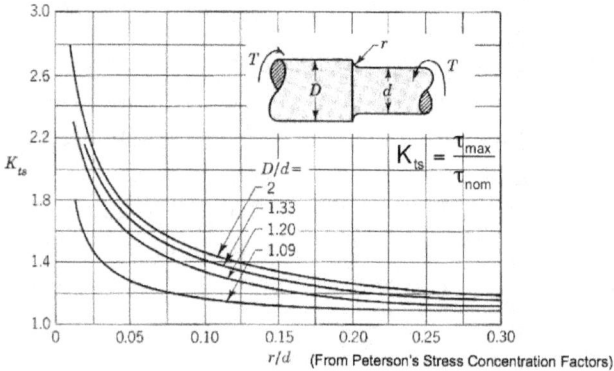

Figure 4-10 Ks for Stepped Shaft in Torsion

Numerical Example-

In this example the numerical variation of K_t for a circular stepped shaft in tension with a diameter ratio of 1.1 is examined as the fillet radius ratio is doubled twice from r/d of 0.025 to 0.050 to 0.10. From Figure 4-8 page 68 this increase in the size of the radius reduces the K_t factor from 2.2 to 1.59 representing a significant decease of 28 percent in the magnitude of the stress at the transition.

Assume an axial load of a circular stepped shaft
d =1.0 in., D =1.10 in., r =.025, .050, .10 in.

then D/d =1.10, r/d = .025, .050, 0.10

and
K_t =2.20 for r/d = 0.025
K_t =1.90 for r/d = 0.050
K_t =1.59 for r/d = 0.10

Figure 4-11 K_t Numerical Comparison

Effect of Material Behavior

The question now is how does the stress concentration effect change with material behavior under load. Consider ductile materials that behave linearly under small strains until yielding takes place after which that undergo significant elongation prior to fracture. In the elastic region the maximum stress σ_m is equal to K_t times the applied load P_1 divided by A (Figure 4-12).

As the stress level in a ductile material reaches the yield condition with P_2 greater than P_1 the material simply begins to elongate which actually relieves the stress concentration effect. In this case the σ_m stress will be less than K_t times P_2 over A. As the fracture point is approached there is no effective stress concentration and the stress is uniform across the section and σ_m is essentially equal to P_3 over A across the entire piece.

Design for Static Mechanical Strength

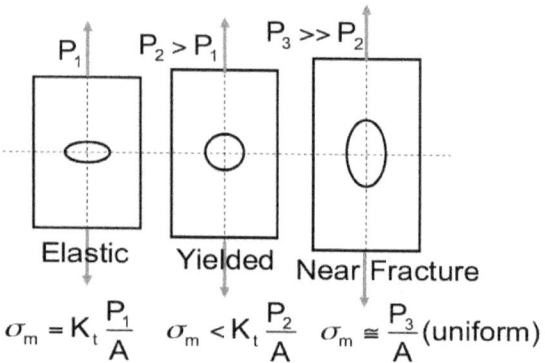

Figure 4-12 Ductile Material Behavior

This relaxation of the stress concentration effect with yielding is nature's way of interjecting its own safety factor into the design.

Since brittle materials normally undergo little elongation prior to fracture there is virtual no relaxation of the stress concentration effect and Kt should be used to determine when fracture will occur. The stress level when fracture initiates will be given by σ_m equal to k_t times P3/A.

Notch Sensitivity and Fatigue

Under conditions of dynamic or fatigue loading the stress concentration factor may vary depending on what is referred to as the "notch" sensitivity of the material's endurance strength. This is accounted for by use of a stress concentration factor K_f that is related to K_t by means of a parameter "q", the notch sensitivity. This sensitivity factor q is determined by comparing the endurance strength obtained with a smooth fatigue test specimen to that of a notched test specimen.

If q is equal to zero the endurance strength is not sensitive to a notch on the fatigue test specimen and K_f is one indicating that a change in geometry has no stress concentration effect. If q is equal to one then the endurance stress is highly sensitive to a notch on the fatigue test specimen and K_f is then equal to K_t the static stress concentration factor.

Sample Notch Sensitivity Chart

Figure 4 - 13 illustrates how the static ultimate strength of steel and the magnitude of a notch on the fatigue test specimen, used to determine the endurance strength, affects the value of the notch sensitivity factor q. As the ultimate static strength of steel increases the value of q increases also. This indicates that under fatigue loading the material behaves more like that of a brittle material under static load requiring the inclusion of a stress concentration effect. For a low ultimate static strength of 60,000 psi and test specimen notch radius of just 0.02 inches the notch sensitivity is about 0.7 which reduces the K_f value, used in fatigue analysis substantially indicating a significant relaxation in the stress concentration effect due to the change in geometry.

Design for Static Mechanical Strength

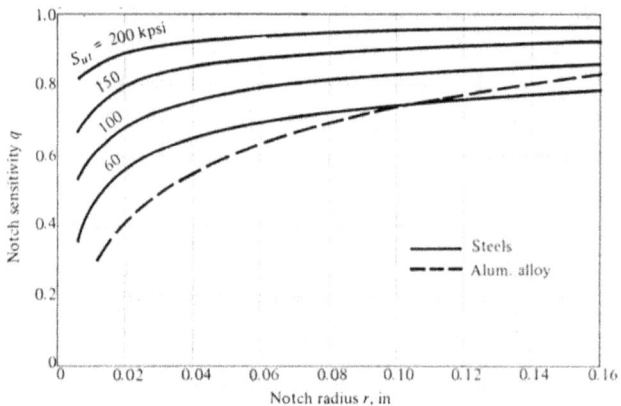

Figure 4-13 Sample Notch Sensitivity Chart

General Guidelines

The general guidelines for including stress concentration effects and values in determining the viability of a given design are summarized in Figure 4 - 14. Under static loading if the material is ductile the stress concentration effect can be neglected unless localized yielding is of concern. For brittle materials under static load the full value of the geometric factor K_t should be used particularly where design failure is based on fracture. For fatigue loading of all materials the stress concentration factor K_f should be used to account for the notch sensitivity testing of the material to determine the endurance limit. If notch sensitivity information is not available then the full value of Kt should be used to give the worst-case scenario.

Design for Static Mechanical Strength

Static Loading
    Ductile Material – neglect stress concentration
    Brittle Material – use geometric factor – K_t
Fatigue Loading
    All Materials – use K_f to account for notch sensitivity of material
                   – if q is not available use K_t to give worst case

Figure 4-14 General Guidelines

The subject of fatigue analysis will be taken up in detail in the monograph Design for Mechanical Fatigue.

Design for Static Mechanical Strength

Design for Static Mechanical Strength

Chapter -5 Material Properties

Chapter 5 deals with how commonly used mechanical strength properties of ferrous and non-ferrous materials are defined and interpreted from experimental tests. Some typical values are included for comparison.

Testing Overview
The two most commonly used mechanical tests to determine the static mechanical strength properties of engineering materials are tensile and compression testing. The tensile test is used to measure standard strength and ductility characteristics of ferrous and non-ferrous metals that significantly elongate under loading. Compression testing is used to determine the strength properties of brittle materials like cast iron, concrete and stone customarily subjected to compressive rather than tensile loading. Fracture under compression normally occurs under limited compressive strain. Additional testing procedures include impact testing to measure energy absorbing properties and non-destructive hardness testing that correlates with tensile strength.

Tensile Test
Figure 5-1 is an illustration of a generic tensile test. The process elongates a test specimen until fracture takes place. The specimen is mounted

between two grips. The lower grip is moved down elongating the specimen at a constant rate. The tensile force generated in the test piece is measured at the fixed upper grip. The amount of elongation is recorded by an extensometer attached to the body of the test specimen.

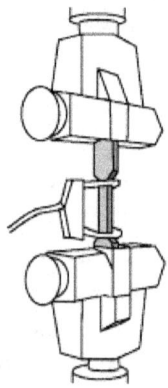

Figure 5-1 Generic Tensile Test

The measured force as a function of elongation is used to determine how the normal tensile stress generated in the specimen behaves with increasing strain. The result is recorded and plotted as a stress versus strain diagram. Figure 5-2 illustrates the classic stress/strain diagram obtained for low carbon steel.

Test Results Interpretation

Initially stress increases linearly with strain. This is referred to as the region of elastic behavior (Hooke's law). It is defined by the slope of the

Design for Static Mechanical Strength

stress strain curve designated the modulus of elasticity or Young's modulus. At a specific value of elevated stress the elongation increases significantly while the stress remains essential constant. This is referred to as the yield point and the associated stress is designated the yield stress or yield strength of the material.

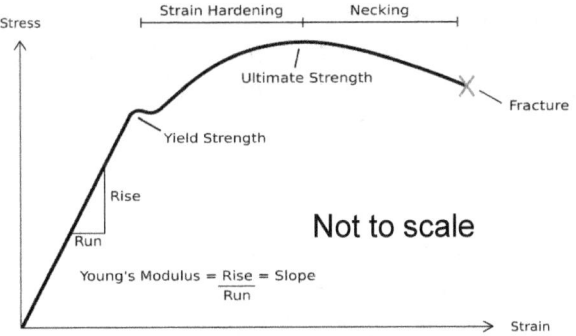

Figure 5–2   Stress/Strain – Low Carbon Steel

Additional increase in strain is accompanied by increasing stress but at a decreased rate. The stress eventually peaks and then begins to decrease with added strain until fracture occurs. From the yield point to the peak stress the material undergoes strain hardening. Beyond that value the cross section of the specimen begin to decrease (called necking). The stress in this region is customarily calculated using the original cross sectional specimen area. This accounts for the stress decreasing beyond the point of ultimate strength. The ultimate strength is designated the tensile strength or the ultimate stress of the material.

Design for Static Mechanical Strength

The area under the curve is representative of the amount of work absorbed by the specimen prior to fracture. The greater this area the more ductility the material possesses. The total percent elongation the material undergoes in the test represents this property.

Carbon steels that have been processed or alloyed along with non-ferrous materials seldom possess a distinct yield point as depicted on the low carbon steel stress strain diagram. For these materials the tensile test results look more like that in Figure 5-3.

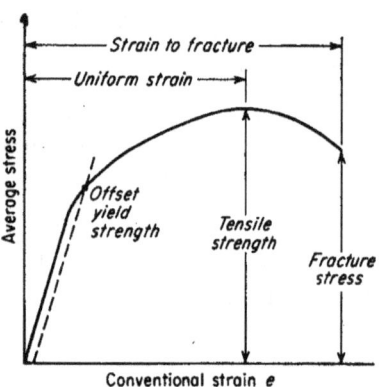

Figure 5-3 Generic Stress Strain Curve

To determine the yield stress for a test result as shown a line is drawn parallel to the elastic behavior at 0.2% strain. The intersection of this line with the test curve defines the yield stress of the material, This is labeled the offset yield strength. Other properties are as previously described.

Design for Static Mechanical Strength

Compression Test

Compression testing of brittle materials is essentially the reverse of a tensile test. A specimen of the material is placed between two platens and a compression strain is applied at a constant rate while the compressive force is measured. A comparison of the characteristic compression stress strain curves for a brittle material compared to the tensile test of a ductile material is shown in Figure 5-4.

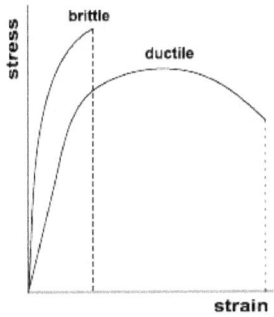

Figure 5-4 Brittle versus Ductile Comparison

It is observed that the ultimate compressive stress may be higher than the tensile strength but that the total elongation prior to fracture is significantly less than that of the ductile material. The areas under the curves to fracture indicate that much more energy can be absorbed by deformation of the ductile material as compared to the brittle material.

Design for Static Mechanical Strength

Sample Properties

Typical ranges of property values for modulus, yield stress, tensile strength and total elongation from tensile tests are presented in Figure 5-5. These include a number of ferrous and nonferrous materials. Individual property values for a specific material of interest can generally be obtained from published test data. Such information may include the effects of mechanical or thermal processing the material has experienced.

	Modulus mpsi	Yield Stress kpsi	Tensile Strength kpsi	% Elongation in 2 in.
Carbon Steel (HR)	28.5 - 30	26 - 49	47 - 90	35.0 - 50.0
Stainless Steel (drawn)	28.5 - 30	15 - 150	125 - 195	10.0 - 20.0
Aluminum Alloy (temp)	10.9 - 14	22 - 59	24 - 63	10.0 - 16.0
Aluminum Alloy (cast)	10.9 - 14	18 - 30	27 - 34	2.0 - 2.0
Copper Based Alloy	14 - 16	50 - 63	56 - 76	5.0 - 8.0
Magnesium Alloy	6.5 - 6.9	30 - 38	40 - 51	7.0 - 9.0

Figure 5–5 Sample Material Properties

Carbon Steel

The effects of alloying and processing on material properties can be dramatic and are well demonstrated by considering plain carbon steel. The effect of changes in the carbon content in hot rolled steel is illustrated in the series of stress strain diagrams in Figure 5-6. As the percentage of carbon by weight, which is small, is increased the yield and tensile stress are both increased quite dramatically. However, the elongation to fracture is significantly decreased. This significant decrease in ductility is indicated by the change in elongation to fracture. At a low carbon percentage the material possesses a great deal of ductility whereas at high carbon levels the material is "stronger" but is more brittle in behavior. Even more dramatic changes occur from hot rolled steels to cold drawn steels of similar alloy composition. The work hardening dramatically improves the yield stress with less but still positive increase in the tensile strength also. However, the effect of cold work processing decreases the ductility for the same composition. Not all material undergo as dramatic changes as carbon steel but it is important to know what the specific composition and properties of a given material are for a given design application.

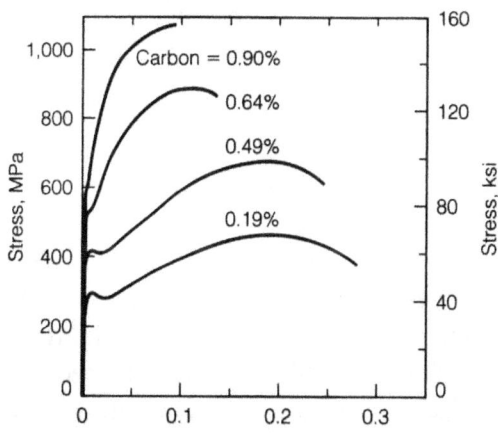

Figure 5-6 Effect of Carbon in Steel

Design for Static Mechanical Strength

Impact Testing

Impact testing is a high strain rate destructive test used to measure the amount of energy absorbed by a material at fracture. This is referred to as a measure of the toughness of the material. The test consists of permitting a weighted pendulum to strike a notched test specimen with a specified amount of kinetic energy. The hammer at the end of the pendulum imparts a state of bending in the sample that produces fracture. The decrease in kinetic energy of the pendulum measured by its height following impact is the energy absorbed by the sample in fracturing.

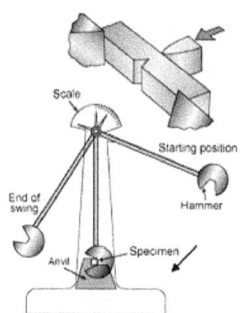

Figure 5-7 Impact Test

Two geometries of test specimen are used. The Charpy test employees the three point bending arrangement depicted in Figure 5-7. The Izod test imparts bending to a cantilever notched specimen held at one end in a vertical position. Both test are used to study temperature-dependent ductile-brittle transition in metals and non-metals. The two tests are e linearly related to one another.

Design for Static Mechanical Strength

Hardness Testing

Hardness testing was earlier cited as a non-destructive technique for measuring static material properties. It is a test that derives from the resistance of material to some form of surface indentation. The two tests commonly used are the Brinell and Rockwell hardness tests. These tests can be related to one another quantitatively and both correlate to the tensile strength of the material. Being non-destructive they have no effect on the strength of finished mechanical parts and can be used to control the quality and uniformity of their manufacture.

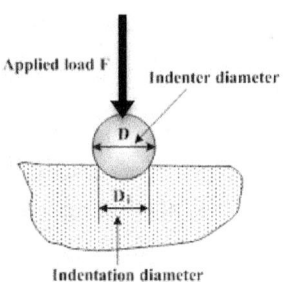

Figure 5-8 Brinell Hardness Test

Brinell Hardness Test

The Brinell hardness test makes use of a loading device that indents the surface of the material with a small 10 mm steel ball under a standard static load (Figure 5-8). The load is removed and the diameter of the indentation is measured with a microscope. The pressure in kg per mm squared is then determined from the surface area of the impression. This pressure is designated the Brinell Hardness number or Bhn. For steels this number generally ranges from 100 to 500.

Rockwell Hardness Testing

Rockwell harness testing also indents the surface of the material but its process and measurements are quite different from the Brinell test. In the Rockwell test the surface is again indented under a standard load. The load is then increased a specified amount and reduced back to the initial load. The increase in the depth of the indentation between the two loads is measured. This increment of change provides the Rockwell harness number. There are two scales of Rockwell hardness. The C scale uses a sphero-conical indenter while the B scale uses a 1/16 in. diameter ball. Typical numbers for medium carbon steel are Rockwell C-20 or B-100 while for very hard steels it would be Rockwell C-100.

Bhn	Rockwell C scale	Rockwell B scale	Tensile Strength kpsi
495	50		247
401	42		196
302	32	107	146
202	15	94	99
149		81	75
101		60	52

Figure 5-8 Brinell Rockwell Comparison

Design for Static Mechanical Strength

The table in Figurer 5-8 lists corresponding example values of Brinell hardness number, Rockwell C scale and B scale numbers together with tensile strength in kpsi for steel. Rockwell C scale measurements are discontinued at 15 because lower values are difficult to obtain and are inaccurate. Rockwell B values are discontinued above 100 for the same reason. The Brinell numbers range from about 100 to 500 for steel. Hardness comparison scales between Brinell and Rockwell numbers for seels are available in sources such as Machinist's Handbook. Good sources for the correlation of these hardness numbers to tensile strength for other materials are more difficult to come by. Such a correlation may need to be determined for a specific material design application by conducting both the tensile tests and hardness tests on that specific material.

Design for Static Mechanical Strength

Design for Static Mechanical Strength

Chapter 6 – Metal Processing

Chapter 6 covers the common forms of mechanical and thermal metal processing and their effects on the strength, ductility and hardness properties of these materials.

Processing Overview

All metals undergo some form of processing in becoming a finished part of a product. This impacts their strength, ductility and hardness properties. These processes fall into two general categories, mechanical processing that changes material properties by physical deformation and thermal processes that produce material transformations in grain size and crystal formation.

Mechanical processing includes casting, the solidification of a molten metal in a mold, hot working that results in large plastic deformations and cold working used primarily to achieve dimensional control and surface quality.

Thermal processing consists of raising the temperature of the material to elevated levels and then controlling the rate of cooling. This includes annealing, quenching, tempering and case hardening. This achieves specific strength and ductility properties in carbon steels and alloys. The effects of metal removal by cutting, machining, drilling and are not covered in this overview.

Design for Static Mechanical Strength

Sand Casting

Casting is simply defined as the pouring of a molten metal into a mold and allowing solidification. The oldest and probably best known of these processes is sand-casting. The mold is formed by packing moist sand around a pattern of the shape to be made. The mold is made of an upper and lower half, the cope and drag.

The pattern (wood, plaster or metal) is made in two halves. If the shape contains an internal cavity a separate core is placed inside the mold before pouring takes place. Vents and risers allow gases to be vented. The metal is poured into the mold through a channel called the sprue.

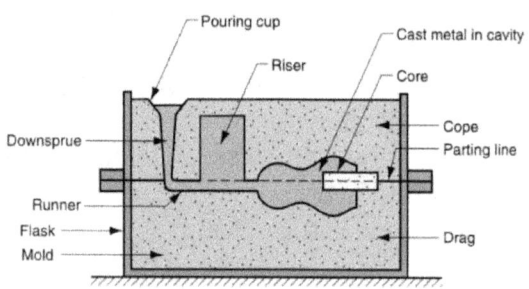

Figure 6-1 Generic Sand Casting

After solidification the mold halves are separated, the casting is removed and the sand is reused. Casting dimensions are nominal and the surface is rough. Sand casting is simple and low cost, used for large parts produced in small numbers.

Design for Static Mechanical Strength

Shown in Figure 6-2 is the cope and drag of a mold that has been separated. The depressions in the sand for the part to be molded along with the pouring sprues, vents and cross runners are visible. Also shown is an automotive intake manifold for a V-8 engine illustrating how complex a casting requiring many internal core elements can be produced by this process.

Figure 6-2 Cope and Drag with a Complex Casting

Investment Casting

Investment casting is used to produce identical modest size cast parts in large numbers to close finished dimensions with a better finish than sand casting. It is often referred to as the "lost wax process".

The part pattern is first machined in metal. This pattern mold is then used to cast multiple numbers of part shapes in wax. These wax parts are then joined by runners to a central tree. The tree and runners become the channels through which the molten metal will flow in the mold when the final casting is made. The tree with the attached wax part shapes is then coated with a slurry of plaster or fine silica. The coated assembly is then baked in an oven during which the wax is driven off leaving a mold of the part shapes, runners and tree.

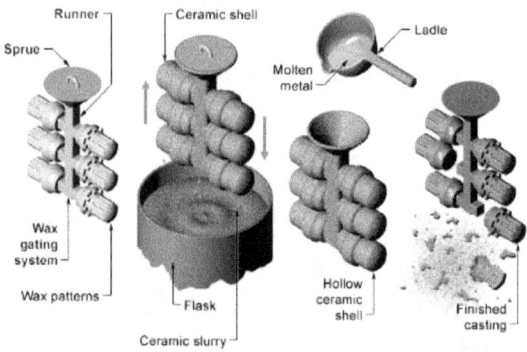

Figure 6-3 Investment Casting Process

The molten metal is then poured into the mold creating multiple cast parts. After solidification the mold is destroyed when it is stripped off the casting. The parts are then separated from the tree. The process is then repeated starting with the casting of the wax patterns of the shape desired.

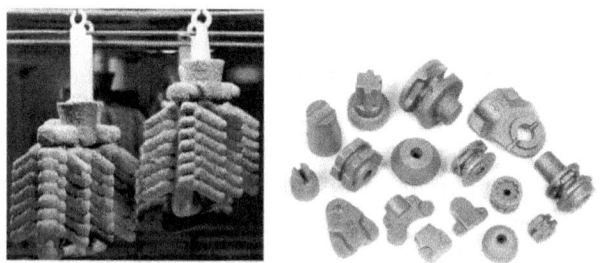

Figure 6-4 F1nal Mold Tree and Sample Parts

Design for Static Mechanical Strength

Shown in Figure 6-4 is a final mold after the wax parts have been attached by runners to the tree and the assembly has gone through the ceramic slurry and baked to remove the wax. Also shown are examples of the types of modest size parts that are well suited to be produced by the investment casting process.

Shell Casting

This process takes its name from the mold used to produce the final casting. A hot metal pattern of machined aluminum, brass or cast iron is immersed in a mixture of sand and thermosetting resin. The heat of the pattern melts the plastic that combines with the sand to form a 3/8 to half-inch shell around the pattern. The system is cured at about 600 degrees F.

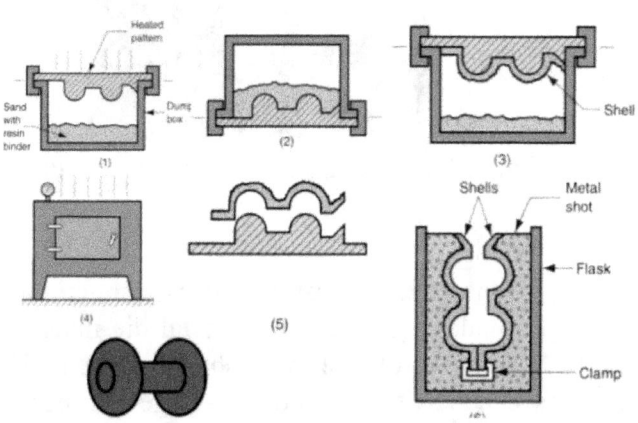

Figure 6-5 Shell Casting Process

Design for Static Mechanical Strength

After cooling the shell is cut in half and stripped from the pattern. The shell halves are then fastened together to form the final mold. The mold is placed in a container of back up material like lead shot to provide support. When the molten metal is poured in the mold the heat burns away the plastic bond allowing the gases to escape and the casting to be air-cooled. This process can achieve close dimensional tolerances and smooth finishes.

Split shell molds for a finned compressor cylinder along with a final cast product is shown in Figure 6-6.

Figure 6-6 Split Shell Mold with Cast Part

Die Casting

In the cold die casting process molten metal is forced under pressure into a metal die machined to be the mold of the part shape desired. This process can be used to produce large numbers of small to moderate sized parts due to the long life of the metal die.

Design for Static Mechanical Strength

In the hot chamber version the pressurizing system is immersed in the molten metal. Chemical reaction between the plunger and molten metal restricts this process to zinc, lead and tin alloys.

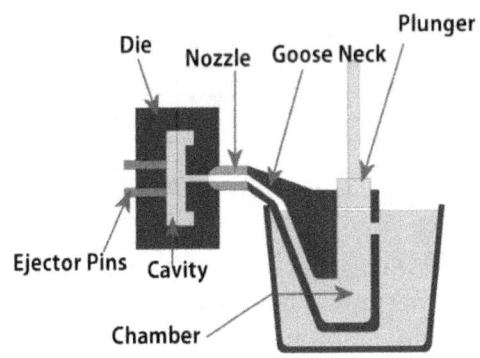

Figure 6-7 Generic Die Casting Process

Aluminum and copper alloys casting are produced in the cold chamber version in which the pressurizing system is separate from the molten medal. The melting point of the cast materials must lie significantly below that of the steel plunger used for pressurization. Injection pressures normally range from 2500 to 4000 psi. An intricate zinc alloy cover plate shown in Figure 6-8 demonstrates the geometric complexity, dimensional accuracy and surface finish quality that can be achieved in the die casting process.

Design for Static Mechanical Strength

Figure 6-8 Typical Die Casting Part

Rolling (ingot to billet)

All hot working processes take advantage of the material being deformed at high temperature allowing large plastic deformations due to the softened state of the metal. The first of these processes that virtually every metal undergoes is the hot rolling of large ingots into slabs and billets. Rolling is the process of passing a red hot ingot between large parallel rolls which reduces it's thickness and increases it length. The hot rolling forming process is performed on steel, aluminum, copper rand magnesium alloys.

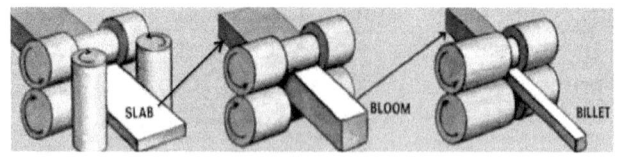

Figure 6-9 Generic Rolling Process

Design for Static Mechanical Strength

By using progressive shaped roll passes as illustrated in Figure 6-9 this process is also used to produce round and square rods as well as structural shapes.

Rotary Piercing

High pressure seamless tubing is produced by passing round rods through inclined rolls that force the rod over a mandrel to form the inside diameter of the tube.. Figure 6-10 illustrates the piercing process along with seamless tubing that has been produced by rotary piercing. It is an example of the type of tubing used in high-pressure boilers n a steam power plant.

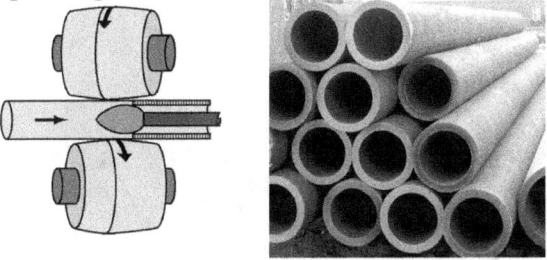

Figure 6-10 Rotry Piercing & High Pressure Pipe

Extrusion

This is a hot working process in which a heated billet is placed in a confining cylinder and forced by a ram through a die that produces long bars with a specific cross sectional shape

Design for Static Mechanical Strength

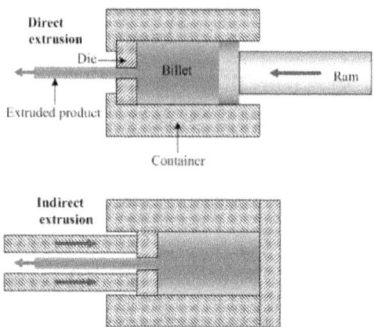

Figure 6-11 Generic Extrusion Process

The process works best on lower melting point metals like aluminum, copper, magnesium and their alloys that flow easier than steel at comparable elevated temperatures. Product can be formed with high dimensional accuracy into very complex solid and hollow cross sectional shapes.

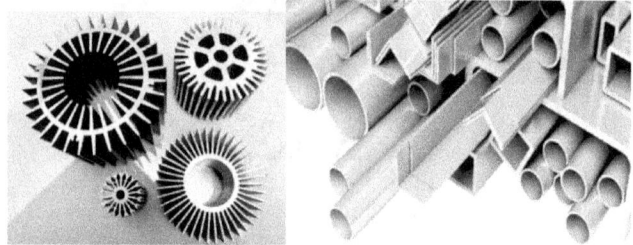

Figure 6-12 Examples of Extrusion Products

It is more costly than rolling because of initial die expense coupled with short life from high wear rates. The process of indirect extrusion moves the die instead a ram, employing hydrostatic pressure around the billet to facilitate metal flow with the use of lubricants to reduce die wear. Figure 6-12 shows

Design for Static Mechanical Strength

extruded product examples of structural shapes as well as short portions of highly complex finned elements.

Forging

In forging large plastic deformations are produced by compressive stresses created by shaped or flat dies using large hammers or presses. This process was practiced by the village black smith with his strong arm and heavy hammer on iron bars heated in an open charcoal furnace aided by hand operated bellows.

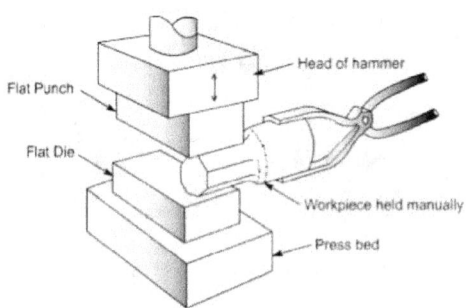

Figure 6-13 Generic Forging Process

This process produces a refined grain structure resulting in higher yield strength and greater ductility than is produced by casting. The die can be open or closed and the hammer or press can be pneumatic or hydraulic driven. It is the first part of the forming process for large items like railroad axles and turbine rotors.

Design for Static Mechanical Strength

Drop forging is a closed die process in which the hammer is actuated by gravity. Closed die forging generally produces a dimensionally more accurate and stronger part than casting.

Figure 6-14 illustrates two extremes of the forging process. On the left is a large ingot being deformed in a huge hydraulic press by a mechanical manipulator. On the right forging is taking place by physically hammering on a hot metal bar against an iron anvil.

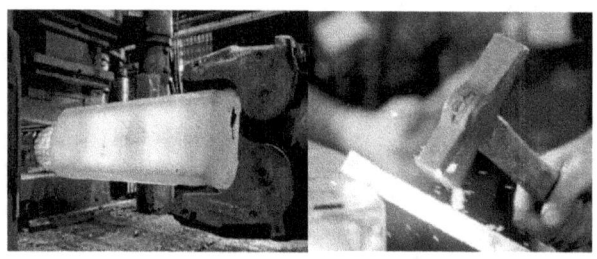

Figure 6-14 Extreme Forging Examples

Cold and Hot Working Compared –

Cold working processes deform metals plastically at or slightly above room temperature. Since the material is less plastic at these temperatures compared to hot working the rates of size reduction are significantly smaller. The grain size of the metal is not appreciably changed however it does become quite distorted. Cold working essentially work hardens the material

Design for Static Mechanical Strength

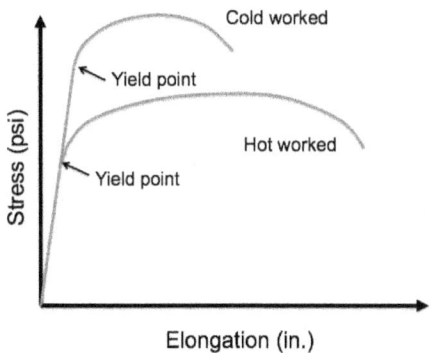

Figure 6-15 Hot and Cold Worked Comparison

producing both higher yield stress and ultimate strength but at the expense of significant reduction in ductility. A typical comparison of properties is illustrated in the graph. The product has a much smoother finish, is dimensionally more accurate and requires less machining or other mechanical finishing. Yield strengths are higher than comparable cast products.

Some Cold Working Processes

Cold rolling is used to reduce hot rolled strip into thin sheet product by passing the material through multiple sets of parallel rolls. Each pass reduces the thickness by some small amount until the final desired thickness is achieved.

Cold drawing is used to reduce the cross sectional dimensions of hot rolled bars. The bars are pulled through dies that can reduce the crosssection by up to 10%. The product is normally

Design for Static Mechanical Strength

referred to as cold finished bars. Even though the product has a smooth finish no metal has been removed only deformed.

Heading is a cold working process in which a portion of the product is upset or flattening out. Examples of cold-headed products are bolts and rivets and other similar shaped parts.

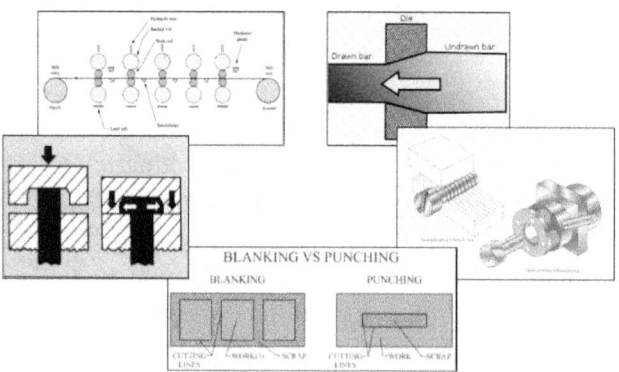

Figure 6-16 Cold Working Process Examples

Roll threading is a cold working process used to form threads on round stock by squeezing and rolling the material between thread shaped dies.

Stamping is a generic cold working term that covers press operations that include blanking, coining and forming or bending.

Shown in Figure 6-16 are the five cold working processes described: cold rolling, cold drawing, heading, roll threading and stamping.

Design for Static Mechanical Strength

Heat Treatment

The heat treatment of metals refers to thermal processing of heating and cooling that interrupts or changes the transformation process of a metal or its alloy as prescribed by its equilibrium phase diagram. These changes can have major effects on mechanical properties.

The more important processes together with their effect on strength, ductility and hardness include annealing, quenching, tempering, and case hardening. The principal difference between these processes is the temperatures to which the material is elevated, how long it may be held at a specified temperature and the rate at which it is cooled. In the material that follows thermal processing of steels will be emphasized.

Annealing

Hot and cold working of ferrous materials introduces residual stresses into the product as well as increasing its hardness or the resistance of its surface to indentation. Annealing is a process that permits the material to slowly transform in accordance with its equilibrium diagram to relieve these two work effects. The temperature of the material is raised to about 100 °F above its critical or equilibrium crystallization temperature and held at that point to allow the carbon in the steel to dissolve.

The material is then allowed to cool very slowly usually in the furnace in which it was heated and held. The process can take hours or days. This softens the material, reduces its hardness, relieves the residual stresses and refines the grain structure.

Figure 6-17 Annealing – Slow Cooling in Furnace

Normalizing –

Normalizing is considered as being included in the annealing thermal processing category. The temperature is raised slightly higher than for a full anneal. The material is then cooled in still air at room temperature. This results in a faster temperature decrease than in a full anneal with subssequently less time for the material to reach its normal equilibrium state. The result is material with a slightly coarser grain size that is easier to machine.

Design for Static Mechanical Strength

Figure 6-18 Normalizing at Room Temperature

This is particularly true in the low carbon steel family, 0.1% to 1.8% carbon content. The material is also slightly harder than fully annealed steel. Normalizing is often used as the final processing treatment.

Quenching

Quenching refers to the rapid cooling of a heated material in some liquid that quickly conduct thermal energy from the metal. The rate of thermal extraction is a function of the quenching media. Two liquids used to quench steel are oil and water.
The resulting structure of the material becomes a function of the carbon content of the steel, the temperature to which the material is heated, the time it is held at the temperature and the rate at which it is then cooled. This time and temperature dictate the degree and type of transformation and crystal structure that will result from the process. If

steel is cooled very rapidly the transformation that takes place produces a structure called martensite. This is a very hard and brittle steel that contains high levels of residual stress.

Figure 6-19 Quenching

Tempering

Steels that have been fully hardened by quenching are very brittle and contain significant residual stresses. A thermal process called tempering can relieve these undesirable conditions. It is sometimes referred to as stress relieving and/or softening. The material is reheated to below the critical temperature and allowed to cool in air. The key element is that the critical temperature is not exceeded. The temperature is dependent on the composition of the steel and degree of hardness desired. The process releases carbon in the martensite that forms carbide crystals. The resulting structure is tempered martensite.

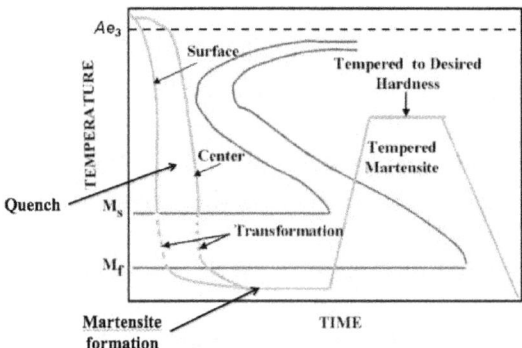

Figure 6-20 Tempering Process

Effects of Tempering

Typical effects of tempering temperature on the more important mechanical properties of medium carbon steels are illustrated in Figure 6-21.

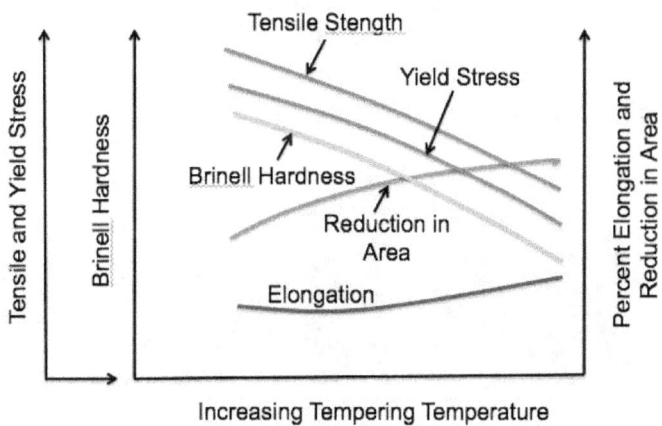

Figure 6-21 Effects of Tempering Temperatures

The yield stress, ultimate strength and Brinell hardness are all reduced as the tempering temperature is increased. The upper temperature is about 150 degrees below the critical temperature where transformation of the structure begins to take place. At the same time reduction in area and the elongation increases with tempering temperature at a somewhat slower rate. Hence the general effect of tempering is to decrease the strength and hardness of the material while increasing its ductility without severely changing it internal structure.

Case Hardening

Case hardening is a process used to preserve the ductility and toughness of the core of a steel part while producing a very hard thin surface finish. This is accomplished by increasing the carbon content of the surface. To accomplish this the steel part is placed in a carburizing media, which can be either a gas, solid, or liquid at some specified temperature for a prescribed period of time.

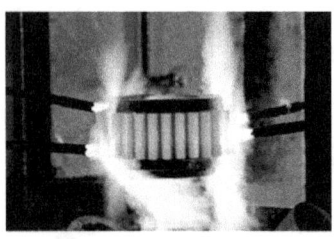 

Figure 6-22 Case Hardening and Product

Design for Static Mechanical Strength

The time and temperature depend on the composition of the steel and the depth of hardness desired. The surface simply absorbs the excess carbon from its environment. The steel is then quenched and tempered. Some of the more common practices include pack carburizing, gas carburizing and nitriding as well as induction hardening and flame hardening.

Design for Static Mechanical Strength

www.ingramcontent.com/pod-product-compliance
Lightning Source LLC
Chambersburg PA
CBHW070300190526
45169CB00001B/490